T0132866

TECHNO-FIXERS

TECHNO-FIXERS

Origins and Implications
of Technological Faith

SEAN F. JOHNSTON

McGill-Queen's University Press

Montreal & Kingston · London · Chicago

© McGill-Queen's University Press 2020

ISBN 978-0-2280-0132-4 (cloth)
ISBN 978-0-2280-0204-8 (ePDF)
ISBN 978-0-2280-0205-5 (ePUB)

Legal deposit first quarter 2020
Bibliothèque nationale du Québec

Printed in Canada on acid-free paper that is 100% ancient forest free
(100% post-consumer recycled), processed chlorine free

This book has been published with the help of a grant from the Canadian
Federation for the Humanities and Social Sciences, through the Awards
to Scholarly Publications Program, using funds provided by the Social
Sciences and Humanities Research Council of Canada.

Funded by the   Financé par le
Government   gouvernement
of Canada   du Canada   Canada   Canada Council   Conseil des arts
                                                        for the Arts   du Canada

We acknowledge the support of the Canada Council for the Arts.

Nous remercions le Conseil des arts du Canada de son soutien.

Library and Archives Canada Cataloguing in Publication

Title: Techno-fixers : origins and implications of technological faith /
    Sean F. Johnston.
Names: Johnston, Sean, 1956- author.
Description: Includes bibliographical references and index.
Identifiers: Canadiana (print) 20190216581 | Canadiana (ebook)
    20190216603 | ISBN 9780228001324 (hardcover) | ISBN 9780228002048
    (ePDF) | ISBN 9780228002055 (ePUB)
Subjects: LCSH: Technology—Social aspects.
Classification: LCC T14.5 .J65 2020 | DDC 303.48/3—dc23

Dedicated to my family
and parents Harold (1930–1985) and Frances (1932–2018)

Contents

Tables and Figures

Tables

Figures

Preface

This is the story of a seductive idea and its sobering consequences.

The twentieth century brought a new cultural confidence in the social powers of invention – along with consumerism, world wars, globalization, and human-generated climate change. This book traces how passive optimism and active manipulations were linked to our growing trust in technological innovation. It pursues the evolving idea through engineering hubris, radical utopian movements, science fiction fanzines, policy-maker soundbites, corporate marketing, and consumer culture. It explores how evangelists of technological fixes have proselytized their faith and critically examines the examples and products of their followers.

The new technological confidence mixed together beliefs that were simultaneously compelling and unsettling. As motor vehicles, electricity supplies, and radio became part of modern life in the early decades of the century, it was hard to argue against the transformative effects and inevitability of such transitions. Like it or not, social consequences seemed to come inexorably with the Machine Age, the Space Age, and the Information Age. This deterministic vision implied an ever more technological future with unavoidable social consequences.

For many, innovative technologies promised appealing new lifestyles and powers. But for a narrower band of proponents – the first generation of technological fixers – wise engineering invention was touted as a guaranteed route to positive human benefits and societal progress. Socially engaged engineers and designers argued that such improvements could be directed, hastened, and amplified.

These engineering adventurers argued that modern societies could be guided only by rational designers. They contended that clever technological solutions could solve contemporary problems better than any traditional method, including economic initiatives, citizen education, political ideology, lifestyle changes, legal frameworks, and moral guidance.

Their seductive claims were tamed by more mainstream American enthusiasts and were eventually boiled down to the concept of the "technological fix." Their shared confidence infused policy-makers, broadcasters, and science popularizers. Trust in technological fixes shaped a new generation of managers and lawmakers, engineers and educators, futurists and citizens – and continues to drive a new generation of techno-fixers today. This cultural wave, its promoters and detractors, have championed the promise and voiced concerns that we have inherited, still unresolved.

This book tracks the hubristic influencers and weighs up the confidences and concerns associated with them: the dramatic potential for novel technologies to work alongside longer human traditions to meet our enduring ambitions – or to reshape society for the worse.

SEAN JOHNSTON, Dumfries, Scotland

Acknowledgments

I thank the archivists responsible for relevant historical collections, notably Margaret Allard for helpful access to the previously unexamined Alvin Weinberg Papers at the Children's Museum of Oak Ridge, USA; Alicia Odeen and Jim Franks for access to the Technocracy Fonds at the University of Alberta Archives, Edmonton, Canada, and at the University of British Columbia Archives, Vancouver, Canada; Lara Michels for preparing the Richard L. Meier Papers and finding aid at the Bancroft Library, University of California, Berkeley, USA; and Kris Bronstad, the archivist responsible for the Weinberg Papers at the Modern Political Archives, Howard H. Baker Jr Center for Public Policy, University of Tennessee, Knoxville, USA.

Valued information providers have included George Wright, director of Technocracy Inc.; John R. Waggener for images from the M. King Hubbert Papers at the American Heritage Center of the University of Wyoming; Mason Inman, Chris Kuykendall, and Ron Swenson for archival information relating to Hubbert; the family of Richard L. Meier, especially Karen Reeds and his former student Eugene Tssui (Lufeng, China); Timothy J. Gawne (Oak Ridge National Lab) and former colleagues Tom Row, Steve Stow, and Alex Zucker on the career of Alvin Weinberg (Oak Ridge, USA). For wider-ranging thoughts on technological confidence, I thank Michael Bove (Massachusetts Institute of Technology, Cambridge, USA), former *Tomorrow's World* presenter Adam Hart-Davis (Devon, UK), and colleagues Benjamin Franks, Ralph Jessop, and Colin McInnis (University of Glasgow, UK).

I am grateful to my commissioning editor, Richard Baggaley, and staff at McGill-Queen's University Press for their enthusiasm and support. I thank journal editors for permission to draw on my publications in *History and Technology*, IEEE *Technology and Society*, *Annals of Science*, and *Technology and Culture*, and to conference attendees, referees, readers and reviewers who engaged with the ongoing work.[1] Students participating in two of my upper-level undergraduate and master's courses, "Health and Technology" and "Environment, Technology and Society," also helped shape the evolution of this text.

A portion of this research was funded by British Academy grant number SG132088.

People say you can't change human nature. We of the engineering profession approach it in another way. The only method of regulating has been to prohibit. You have noticed the sign, "Passengers are prohibited from standing on platforms," in railway cars. Engineers came along and designed a train without platforms and said, "Stand on them if you can."
   Howard Scott, public lecture, Winnipeg, 1937

Social problems … deal primarily with interrelationships between human beings, and do not now involve materials or the man-machine interaction. Most people would instinctively exclude the scientist and the technologist in the search for solutions. Yet, in many cases, a social problem can be restated so that it is also a scientific or an engineering problem that is not only researchable, but soluble!
   Richard L. Meier, *Science and Economic Development: New Patterns of Living*, 1956

The technologist – i.e. the real engineer – has much to offer in the solution of social problems that are usually considered to be the province of those who try to *manipulate* social behavior; it is the latter whom I call "social engineers."
   Alvin Weinberg, letter to Waldo E. Smith, 1967

My bent is … an engineer's bias, which sees everything in terms of solvable design problems   Engineers figure out what the problem is, and then make it go away.
   Stewart Brand, *Whole Earth Discipline*, 2009

There's an app for that!
   Apple Corporation, 2010

1 Modern Times and Technological Faith

Introduction: What Needs Fixing?

In 1966, a well-connected American engineer posed a provocative question: Will technology solve all the problems of modern society? He argued that it would, and soon. Even more contentiously, he suggested that rational designers could eventually supplant social scientists – and perhaps even policy-makers, lawmakers, and educators – as the best trouble-shooters and problem-solvers for society's ills.[1]

The engineer was the director of Tennessee's Oak Ridge National Laboratory, Dr Alvin Weinberg. An active essayist, networker, and contributor to government committees on science and technology, he was to reach wide audiences over four decades. Weinberg did not invent the seductive idea, but he gave it a memorable name: the "technological fix." Its bold claim is that *technological inventiveness can reliably bypass traditional approaches –* such as education, political governance, moral guidance, and laws – *to address social problems.*

Weinberg's notion echoed and clarified the views of his predecessors and contemporaries. His rhetoric was a call-to-arms for engineers, technologists, and designers, especially those who saw themselves as having a professional responsibility to improve human welfare. It was also aimed at institutions, offering goals and methods for government think tanks and motivating corporate mission statements.[2] The rhetoric of technological fixes is now implicit in modern entrepreneurial culture. Indeed, today the term "innovation" is trumpeted by industrialists, policy-makers, and university administrators as synonymous with economic competitiveness, national flourishing, and societal beneficence.[3]

But the idea is a slippery one. Initially promoted by a fringe of radical engineers, its allure draws on a constellation of more widely subscribed cultural beliefs and values. At the turn of the twentieth century, growing audiences in modern countries, and especially North America, became captivated by new inventions and their implications. Many came to view technological change as inevitably life-transforming and unstoppable. A smaller fraction saw new technologies as tantamount to societal progress. A still smaller subset nuanced the optimism further: with wise technical designers at the helm, they argued, society's problems could be eliminated one by one. The wonder of the technological fix is that it became orthodox and so widely shared during the past century. It translated passive popular confidences into active manipulations by zealous engineers. A key nuance, seldom explicit, was the *intentionality* of wise designers aiming to create technologies to achieve social ends. In distinct flavours, trust in technological fixes was echoed by utopian writers, popular media, and consumer culture. Our attraction to technological solutions for the problems of daily life is a key feature of contemporary lifestyles and dreams of the future.

Yet this pared-down sectarian faith has proven notoriously difficult to tame. Weinberg's own promotion vacillated from tentative to hubristic. Inventive solutions to societal problems, he claimed, may range from specific "quick fixes" to enduring social transformations. Enthusiasts have praised the self-evident validity of technological fixes, while critics have highlighted their naïve assumptions and simplistic doctrine about how technology and society interact. In fact, entire professions and academic disciplines line up against each other to promote or condemn technological fixes.

Techno-Fixers recounts, unpicks, and challenges these often amorphous and untested ideas, and shows how their promoters evolved from marginal thinkers to mainstream influencers confident about technological solutions to human problems. It unwraps Weinberg's problematic package, identifies the origins and evidence for its claims, follows its optimistic supporters and contentious outcomes, and explores the implications for present-day technologists and citizens.

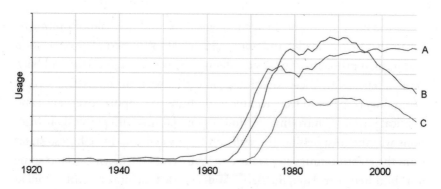

Figure 1.1. The rhetoric of modern problem-solving. Global English usage of the terms (A) "technological solution," (B) "technological fix," and (C) "technical fix," according to Google *n*-gram analysis.

Trusting Fixes: A Prehistory

While describing the core notion of the technological fix is easy enough, tracing its sources and circulation throughout culture is more difficult. One way to begin tracking the currency of a concept is via phrase-usage statistics. The invention and popularity of new terms can reveal new topics and discourse. The "*n*-gram" is a useful tool that analyzes a large range of published texts to determine frequency of word usage over time for various languages and dialects.[4] Variants of the term "technological fix" emerge during the 1960s (figure 1.1).

We can trace this across languages. In German, the term "technological fix" has had limited usage as an untranslated English import and is much less common than the generic phrase *technische Lösung* ("technical solution"), which gained ground from the 1840s. In French, too, there is no direct equivalent, but the phrase "solution technique" broadly parallels German and English usage over the same period. And in British English, the terms "technological fix" and "technical fix" appear at about the same time as American uptake but grow more slowly in popularity. So, usage hints that there are distinct cultural contexts and understandings for these seemingly similar terms. Its varying currency suggests that the term "technological fix" became a cultural export popularized by Alvin Weinberg's writings on the topic but that it was related to earlier discourse about technology-inspired solutions to human problems.

Yet the prescription of the technological fix, and the halo of optimistic beliefs that surrounded it, did not originate with Weinberg. The conviction that human problems can be side-stepped by clever engineering had been applied repeatedly, if less consciously, in earlier societal contexts. The most commonly recognized were technical solutions to support governance and political power.[5] Controlling the flow of populations with physical obstructions was the idea behind the walls built between warring Chinese states from the eighth century BC, Hadrian's Wall across northern England some nine hundred years later, the Berlin Wall (1961), and, most recently, Donald Trump's Mexican wall. Such barriers were conceived to supplement or entirely bypass other methods, such as diplomatic negotiation, laws, policing, or social accommodation.

Other military technologies can amplify these social powers. Rather than passively protecting, as walls do, they may provide the means to threaten or harm, and to actively defend or extend territories of control. Culture by culture and decade by decade, military tacticians have sought to exploit new inventions to their advantage in waging war.[6] It is significant that, while the most successful of them have generally recognized the close association between styles of combat and military hardware, their respective leaders and funders have more often identified *matériel* rather than personnel as the essential element in military success.[7] This simplification has tended to overlook human tactics, organization, and skills or, alternatively, has reconfigured the human elements as components of a military machine. This translation of human factors into technological elements is at the heart of technological fixes.[8]

Such "fixes" for situations of conflict often seem intuitively appropriate. When social methods – intellectual, legal, moral, and educational – are inadequate, the obvious fallbacks are more primal physical constraints. Artefacts serve as proxies for other human methods that are judged too slow or too complex. Conceived in this way, technological fixes are potent responses to immediate problems.

Yet there is an equally long history of positive technological solutions. Since at least the Neolithic period beginning some ten thousand years ago, human cultures have increasingly recognized the close affinity between technologies and human prospering. Admittedly this is a skewed perspective; the modern archaeological narrative is dominated by cultures of tool-making because artefacts are more durable than are the traces of social life. Technologies of stone, copper, and bronze – each of which involved sophisticated sociotechnical systems of mining, trading, craft skills, and contexts of usage – were successively exchanged for iron-based expertise. From our modern perspective, we can presume that users of improved tools valued the social side effects of efficacy and labour saving.[9] Similarly, the invention of agriculture (selecting, planting, and harvesting, and later irrigating and fertilizing crops) went hand-in-hand with rising populations (even if the direction of influence is unclear in the archaeological record).[10] Adoption of aqueducts and human-actuated pumps allowed some population centres to expand and, in so doing, to acquire greater political power over neighbours and to improve social stability.[11] We can infer that these technological transitions carried perceived benefits, at least for the most influential members of those societies. Thus, as argued throughout *Techno-Fixers*, trust in technological fixes is a cultural perception having profound implications for human actions and power relations.

Technological fixes have also been common where social solutions are difficult to discern or seemingly impossible to implement. Prosthetics for cosmetic enhancement, for example, have been widely employed through human history to compensate for perceived bodily disfigurement. Artificial toes of wood and leather were used in ancient Egypt; artificial eyes and nonfunctional hands in the ancient and mediaeval world normalized the appearance of afflicted individuals and provided a psychological sense of wholeness, aiding their reintegration into society.[12] Modern examples of add-on bodily technologies include hairpieces, teeth-whitening products, and cosmetics for skin blemishes to address similar cultural issues.

Yet societal adaptations are also conceivable and have been adopted at least partially in some cultures – for instance, the greater social inclusion of amputees.[13] A contemporary illustration is the growing acknowledgment of wheelchair users in modern societies. The transition in recent times has been driven largely by social accommodations: special interest groups highlighting discrimination of opportunity; legislation to extend rights of access to public buildings and employment; growing cultural acceptance of disability and diversity. In such cases, the technological solutions, such as wheelchair ramps and adapted vehicles, are important but subordinate to societal adjustments.[14]

As these examples hint, technological fixes are particularly common for social issues surrounding human well-being and health. Socially defined harms frequently attract technological solutions. For instance, an important domain of health problems is lifestyle-related obesity. Through the late twentieth century, technological remediation became increasingly varied and popular: diet aids to reduce calorie content or to inhibit absorption of fats from foods; gastric bands to constrain consumption; liposuction to remove body fat; coronary procedures to remove blockages or expand arteries. Perhaps the most dramatic of technological fixes for lifestyle and diet-induced illness was the heart transplant, first trialled to public acclaim during the late 1960s and spawning unflagging hopes for artificial hearts.[15] In a period of unprecedented access to inexpensive food, techno-fixers popularized scientific nutrition via over-the-counter vitamin supplements and diet aids.[16] Such solutions, they argued, could correct for unbalanced dietary regimens, low income, hectic lifestyles, inexpert cooking, or lack of will power.[17]

Technological responses to obesity suggest another recurring theme of this book: the frame for identifying problems and the scale of problem-solving. Individual "fixes" for obesity pin responsibility on individuals and may be the preferred solution for those who acknowledge their inability to diet or exercise adequately. But other factors in modern societies also play a role in constraining individual choices. Urban environments may make exercise socially inhibiting or even dangerous; the economics of mass-produced fast foods may encourage nutritionally unbalanced high-calorie diets; marketing and advertising campaigns may misrepresent and over-sell unhealthy foods; peer groups and wider cultural milieus may promote sedentary lifestyles. In the face of such large-scale and systemic social factors, organizing societal responses to obesity may appear to be more difficult. Traditional social methods of behavioural change – such as measures to educate, morally encourage, or legally enforce individual lifestyle changes required to avoid obesity – seem difficult enough. To challenge societal structures (economic, cultural, and political) that shape those individual choices may appear hopeless. In these circumstances, technological solutions often seem ideal. By narrowing the problem to a single specific cause having a technical work-around, they offer a measurable positive outcome. Each of these examples illustrates the principle: when social techniques are judged inadequate for solving problems, technological fixes are devised to detour around them.[18]

A few broad contemporary issues can suggest these prevalent attitudes. A first domain is resolution of acute environmental problems. As environmental concerns rose from the late 1960s with growing attention to air and river pollution, oil-tanker spills, and fears about nuclear waste, technological quick fixes were proposed as timely and reassuring solutions. Current options include oil-digesting microbes to deal with spills and industrial waste, biodegradable packaging, biotechnologies for fuel production, and schemes for addressing anthropogenic climate change via geo-engineering.[19]

A second domain of problems attracting technology-dominated responses from government and industry is terrorism. As airplane hijackings proliferated during the early 1970s and more varied threats were identified after 2000, technologists responded with imaginative solutions ranging from low-tech lockable cockpit doors to body-scanning and contraband-detecting systems to technologies monitoring global internet communications. In the tradition of technological fixes, these single-minded hardware solutions are rapid responses to events that have relatively complex social, political, and/or economic roots.[20]

More recently, software technologies have been embraced by consumers as even more seductive ways to supplement personal skills, improve efficiency, and empower lifestyles – a marketing philosophy dubbed "solutionism" by a handful of advocates and their critics.[21] By sidestepping the traditional societal tools of education, skills development, or political action, such software solutions are technological fixes in precisely the form proselytized by Weinberg.[22]

So, adoption of technological approaches to solve human problems has been a long-standing tactic of societies, and it remains so. Such examples suggest that technological fixes may be conceived not merely as an emergency stopgap but also as a general societal approach. The rapid innovation over the past century has generated widespread confidence in the potential of technologies to improve modern life and society. Indeed, *modernity* can be defined as the outcome and expression of this cultural view.[23]

Aims, Approach, and Audiences

Techno-Fixers traces the evolution of the technological fix as an explicit concept, as a diffuse cultural belief, and as a contentious practice. The early chapters focus on the first prophets of technological problem-solving, who voiced a strategy for social progress founded on science. The middle chapters step back to illustrate the reflections of this optimistic philosophy in popular culture and its expression in real-world engineering projects and products. The final chapters step back yet further to identify generic features, assess implications, and connect the dots in an analytical overview.

This approach juxtaposes scales and disciplines. The locus for much of the coverage is North America, where the most deep-seated technological faith has prospered, with side trips to Europe and, on occasion, Asia. The window of attention is the past century of modernity, when the technological fix was first enunciated as a principle and applied enthusiastically to problem-solving and progress.

The topic begs for interdisciplinary treatment. For over a century, technical workers have been guided by this implicit faith in the positive social power of technologies. Over the same period, social science has gained authority for guiding public policy and corporate behaviours. And shaped by these two powerful societal forces are their more visible products: material production, pragmatic politics, and popular culture. Thus the book's attentions are interdisciplinary. Much of the coverage is historical (although the final chapters link the rhetoric of early radical engineers to present-day technological adventurers, and their twentieth-century practices to twenty-first-century negative outcomes). Part of this historical perspective is informed by the history and philosophy of science (HPS), which relates intellectual history to professional practices. Part, too, will be familiar to social and cultural historians of technology attuned to how technologies are conceived, viewed, and taken up by their users. And another equally important historical and philosophical strand informs the final chapters, where environmental history, social studies, and ethics of technology provide a basis for evaluation. This constellation of perspectives is sometimes summarized by the labels science and technology studies (STS) or science, technology, and society (also with the acronym STS). STS is an outward-looking field that seeks wider audiences: not just academic scholars of history, philosophy, and social studies but, even more importantly, practitioners of science and engineering, social and technology policy, economics, and the wider publics enrolled in their activities.

The intended readership for this book, then, is envisaged not only as academics and students of the social sciences and humanities but also as professionals and educated non-specialists. For each of these audiences, some of the historical actors and episodes may be recognizable but, I hope, rendered less familiar when viewed through the lens of technological fixing. My aim is to trace the half-recognized and seldom-challenged currents of technological faith and to reveal their problematic assumptions.[24]

Overview of Organization and Argument

Prophets of Technological Problem-Solving

Confidence in the transformative social powers of technology was promoted first and most consistently in North America by a handful of self-identified enthusiasts and proselytizers. Centred initially on groups associated with autodidact engineer Howard Scott (1890–1970), the nascent concept was later refined and championed through the careful postwar studies of academic Richard L. Meier (1920–2007) and the speech-making of physicist-administrator Alvin Weinberg (1915–2006) as "the technological fix." The following three chapters trace such discourses about technology from the era of high industrialism between the world wars through the nuclear age and beyond to inform forecasts of the future.

The time frame, historical correlations, and methodology of these early chapters are noteworthy extensions of prior researches. Scott and Weinberg are both well known to historians of the twentieth century in the distinctive contexts of interwar Technocracy and postwar nuclear power, respectively. These figures and their organizations have previously been studied separately and over the periods of their greatest public prominence, and with attention to a more diffuse collection of episodes and themes.[25] Beginning from such segregated accounts, the chapters of this book trace the intersecting activities and messages of key promoters over a century. Their focus is not the brief flowering and decline of a fringe movement or of societal experimentation with innovative sources of energy; rather, they take a long view, tracking the promotion of engineering solutions for societal problems – a notion that resonated with varieties of popular faith in technological fixes.

My research is based, in part, on hitherto unavailable archival holdings that chronicle this broad timespan via unpublished correspondence, speeches, exhibition materials, and limited-circulation texts. These archives consequently provide insights into evolving notions of technological fixes and into the development of influential rhetorical practices.[26] Later chapters examine episodes more familiar to historians and wider publics, but they do so by applying the defamiliarizing lens of technological fixes to connect a broad range of historical actors.

What has made promises of technological fixes so seductive? Attention to the nature of this discourse, and its echoes in wider culture, is at the methodological centre of this comparison of key techno-fixers.[27] Howard Scott, Richard Meier, and Alvin Weinberg took ideas that were circulating in the engineering world and boiled them down, devoting decades to promoting them to distinct audiences. In the early chapters, focusing on a close reading of the speeches, articles, and illustrations employed by both Scott and Weinberg, I argue that the techniques of popularization they adopted were markedly different from traditional engineering communications. This rhetorical focus illustrates how their style of dissemination, as much as their rationale, promoted cultural confidence in the slippery notion of technological fixes.

I argue that techno-fixers presented self-evident examples of inventions for societal ills in the form of easily absorbed tales. They reshaped the radical discourse of interwar technocracy into a style of communication amenable to post–Second World War policy-making and public understandings of science and technology. Richard Meier melded idealism and enthusiasm with caution, providing detailed foundations for how technological fixes might work. By contrast, the key determinants in this transition were the characters of Howard Scott and Alvin Weinberg: these men functioned as energetic missionaries, and the form and content of their rhetoric supported a form of persuasion akin to religious discourse.

Chapter 2 reveals how much of Weinberg's message and rhetoric had been trialled a generation earlier. Journalists after the First World War christened modern culture the "Machine Age," a period that vaunted the mechanization of cities and agriculture, industrial efficiency, "scientific management," and, most of all, an engineering can-do approach to modern problems.[28] Social progress became associated with applied science. Electric appliances, for example, extended productivity and leisure pursuits; radio entertained, educated, and united the nation; motor vehicles and aircraft provided greater personal mobility for at least a privileged few.

But praise of technological change was accompanied by criticisms of the imperfections of modern society, often by the same analysts. Among the most strident voices were members of the group initially called the Technical Alliance and, later, Technocracy Inc. They railed against the problems of waste, inefficiency, and the incompetence of industrialists and government leaders, and called for the application of "the achievements of science to societal and industrial affairs."[29]

As I discuss in Chapter 2, the cadre boiled down ideas circulating among European scientists and designers and distributed them to wider American publics. Their common-sense conviction was that social measures could be rendered unnecessary by wise engineering. While remaining a fringe organization of vacillating popularity, Technocracy Inc. was emblematic as the first group to seek generic links between technology and society. The Technocrats championed a cluster of notions that was to provide foundations for later visions of the technological fix, coining the phrase "technological social design." Their courses, exhibits, parades, and speeches did not explain or label any of these beliefs, but their rhetoric asserted the ideas as unassailable truths. Their popular discourse also prioritized the role of technical experts as modern authorities. Only engineers and technicians, they argued, could manage an industrialized society. Politicians and political solutions were an anachronism to be discarded. Their verdict relied on the idea of "technological determinism," which asserts that technological innovation inevitably transforms society.[30] Between the lines, their message about "technological progressivism" linked such innovation to social improvement. Just as importantly, the Technocrats seeded this faith in wider cultural discourse.

The resonances between expressions of modern technological confidence, social progress, and religious faith had been remarked as early as the 1920s, with Dora Russell, for example, linking American industrial zeal ("the dogmas of machine-worshippers") with the social ideals of the Russian Revolution.[31] Both the ideological and the theological connotations of this conviction, and, more particularly, the style of communication by which it was promoted, are threads that are interwoven throughout this book, which focuses on how this deceptively discrete and simple faith has been proselytized to influence wider cultural creeds. Scott and Weinberg preached tales of wise technological problem-solving to distinct audiences. Technocrats tailored their populist message to disaffected Depression-era North Americans; Weinberg focused on the movers and shakers of the Cold War world. Recounting universalized parables of engineering authority and honed by years of repetition, the sparse narratives and concrete examples attracted receptive audiences. Their enthusiasms continue to be echoed by more recent technological adventurers to inspire investors and wider publics.

Human Conflict and Technological Confidence

As shown in chapter 3, these influences were enduring. Even after the Second World War Technocrats offered appealing explanations to new audiences, which now included new cohorts of freshly trained technologists, scientists, and engineers. For many, wartime experiences had illustrated overwhelming technological progress alongside equally obvious social and ethical problems that needed to be solved (figure 1.2).[32]

A generation of technological optimists applied their new-found confidence to tackle socially complex problems. The work of Richard Meier epitomized their mindset.[33] Postwar urban design, for example, would confront endemic poverty, inadequate housing, and citizen ill-health by attending to their technological dimensions. Proponents would later recommend this pared down rational approach for less-developed countries as the most effective method of reproducing Western standards of living and shaping social norms of well-being.[34]

Postwar planning provided evidence that rationalized housing, transport, and communication networks could quickly improve the quality of life in cities under any political system. Nascent nuclear energy projects also channelled the promise of new technology to transform societies. During President Eisenhower's somewhat cynical Atoms for Peace initiative of the mid-1950s, for instance, atomic energy was forecast as a means of irradiating food to hinder spoilage, desalinating seawater to irrigate deserts and increase agricultural production, and supplying low-cost electrical power to boost economies.[35] For past participants in the Manhattan Project in particular, nuclear energy appeared to offer new social horizons and career orientations: the new technologies were a model of how to apply rapid innovation to recalcitrant human problems that had outlasted the war.

Chapter 4 explores Weinberg's reduction of such initiatives to a concise principle that brought the notion to new audiences of establishment engineers, consultants, and policy-makers. He was to describe the Manhattan

Figure 1.2 Engineers and scientists as social problem-solvers.
From Alvin Weinberg's personal collection.

Project, for instance, as a prototype for grand technological solutions to societal problems. The creation of the atomic bomb, he was to suggest, illustrated how the planned invasion of Japan and negotiation with Russian allies over invasion plans had both been circumvented by a technological solution. Similarly, he credited the threat of the H-bomb as being a technological solution to the problem of war that did not require changing human nature.[36] This chapter illustrates how technological fixes became popular within business and government as solutions to novel and acute problems, and how they have remained so since.

Popular Optimism for Technological Answers

Chapter 5 moves from close examinations of key techno-fixers to an overview of their audiences, tracing how confidence in technological solutions was expressed through wider channels in popular culture. It details how the secular faith in the progressive nature of technologies was not limited to a naïve period of early engagement but grew into an assurance embedded in modern life. Utopian tracts, world's fairs, and science fiction fandom supplied potent catechisms, and the wares of manufacturers carried the promise of heaven on earth. I argue that popular audiences became passive congregations for these sermons of progress.

While such panaceas had attracted relatively powerless audiences before the Second World War, the postwar successes attracted wider social groups, business interests, and media narratives. Major technological projects provided confidence in engineering ingenuity to achieve ambitious goals. The space race addressed seemingly insoluble technical challenges and – as trumpeted by NASA, its contractors, and cooperative media sources – spun off associated technologies for consumer benefit.[37] Urban planners supported regeneration projects in which reconfigured infrastructure would transform social life, such as implementing expressway networks in lock-step with urban renewal. Supporting these enthusiastic forecasts was the widespread but seldom interrogated popular faith linking technological progress and social progress.

This chapter surveys how the resonances in modern culture can be discerned in shared visions of the future and in how contemporary problems have increasingly been framed and addressed in narrow technological terms.[38]

Shortfalls of Inventive Solutions

Yet these inspirational visions have frequently been myopic, focusing on hopes rather than on actual outcomes. Technological fixes are founded on belief in the positive outcomes of intentional design. Chapter 6 traces the life histories of some prominent technological innovations to assess their longer-term effects and side effects, a perspective still lamentably uncommon in engineering analysis.

Like expressions of technological faith, critiques of technology have grown around particular examples. Opponents of the Vietnam War, for instance, cited the impotence of sophisticated military systems against the guerilla methods of a resourceful enemy.[39] If high technology can be negated by such social and political opposition, this seemed to suggest, why should technological fixes be trusted as a cure-all for social and political problems? For urban audiences over the same period, nuclear technologies were increasingly criticized as inherently dangerous. For growing publics, the field represented a failure of government-managed safety certification procedures and a secretive industry. Similarly, the chemical industry, which had once been lauded for technological fixes such as DDT to kill agricultural pests and assure high crop yields, was now criticized as the source of widespread ecological damage.[40] Such technological criticism pointed to catastrophes such as super-tanker spills as representative of decision making that prioritized the global petrochemical economy over regional interests.[41] Identifying beneficiaries and benefactors became increasingly contentious.

Dissenting contemporaries of technological faith have included academics from other fields, weighing in with critical perspectives drawn from sociology, economics, and ethics. Such examinations have nevertheless been uncommon. More typically, the side effects of technological innovation have been experienced by relatively voiceless publics and environments. Historians of technology, the environment, popular culture, and politics, attentive to such voices, have increasingly disparaged the dreams of technological fixers. Chapter 6 argues that retrospective analyses paint a more pessimistic picture than do the forecasts. The historical cases argue that technological solutions have typically streamlined analysis; prioritized particular economic, corporate, or consumer interests rather than wider benefits; and underestimated societal side effects.

Challenging Democratic Society

In chapter 7, I examine such misgivings to identify generic characteristics of technological fixes, refocusing the discussion begun by Weinberg a half-century ago. I focus on the implications for political power and the governance afforded (or denied) by technological fixers. The role of technological experts as decision makers in modern society has been variously configured and contested throughout the past century, from the heavily engineer-weighted Soviet politburo in mid-century to the role of technologically loaded consumerism today.

The faint voices of the beneficiaries – and potentially victims – of technological fixes are of concern. Howard Scott's Technocrats anticipated that engineers would replace inexpert policy-makers, politicians, and economists by a *technate*, or technological government. According to Alvin Weinberg, government-assigned teams of engineers could assume responsibility for addressing social problems for the national good. For Richard Meier, the process of directing technical solutions was envisaged as cooperation between engineers, paternalistic organizations, and communities but ultimately guided by those with expert knowledge. Such management by elites might be assessed and even voted upon by wider audiences, but this consultative process threatened to undermine the special role of technological competence in such a rational society. The effects of public participation in engineering solutions raised mixed feelings for Weinberg, who observed with chagrin that some of his technological solutions were unlikely to succeed in a liberal democracy and that "nuclear energy seems to do best where the underlying political structure is elitist."[42] As explored in this chapter, the same issues may disempower communities or individual consumers who opt for technological fixes. They may fail to identify how the "problem" and "solution" have been framed by the designers, companies, governments, or media sources who promote them. As a result, the solutions they are offered may be shallow, short-term, or off-target, reproducing social inequities and undiscerning cultural values.

This chapter critically assesses technological solutions in the round. Academic critiques have argued that techno-fixers are naïvely complacent about the methods and outcomes of science ("scientistic") and tend to narrowly define complex problems and feasible solutions, a simplification known as "reductionism."[43] Because of its exaggerated attention to observable conditions and measurable outcomes (the philosophical stance of "positivism"), they suggest, this form of rational decision making carries epistemological and ethical weaknesses. This set of confidences necessarily devotes less consideration to aspects of human values that cannot be reduced to measurable quantities.

The focus on outcomes also identifies the link between technological fixes and utilitarian ethics, in which the goal is to maximize positive outcomes ("the greatest good"). This ethical framework works well for purely engineering problems but can disfavour groups or environments that are not identified as intended beneficiaries ("the greatest number"). There are other ethical alternatives for judging responsible innovation: duty-based ethics (deontology) and virtue ethics, which instead focus on rights and on personal behaviours, respectively.[44]

In this chapter I argue that the narrowing of analytical dimensions is particularly dangerous when problem-solving relies on technological fixes: How can we adequately assess whether a solution satisfies the unvoiced or inexpressible wishes of all those affected? The problem becomes acute when we consider communities, environments, and species without a voice.

Philosopher Arne Naess criticized such ethical implications of relying on technological solutions. He argued that popular enthusiasm for such fixes tended to prioritize the status quo, that is, the interests of current ways of life, and especially current socio-economic conditions and beneficiaries. Naess argued that technological fixes carried cultural presuppositions about what was "reasonable" and, consequently, framed problems narrowly. They generally underestimate the scale and nature of sociotechnical problems and the potency and side effects that engineering solutions can offer. Naess's alternative analysis sought to consider social, cultural, and technological solutions in tandem, and identified technological fixes as simplistic and inadequate.[45]

For an even wider range of theorists, the technological fix has been portrayed as hubris, or excessive confidence, regarding human abilities to adequately understand and manage society and nature through rational means. As a "band-aid" solution to problems involving sophisticated systems, technological fixes were argued to both underestimate and inadequately solve complex problems. Philosopher Alan Drengson, for example, explored the moral values and religious underpinnings of these wider critical perspectives.[46] He argued that technological solutions were too often short term and incomplete, and consequently could camouflage the ultimate sources of larger problems and the nature of genuinely satisfactory solutions.

Chapter 7 consequently develops the theme that engineers have important responsibilities regarding technological solutions. They should temper their paternalism and recognize the diverse public as active participants rather than as passive recipients. Designers need to pay close attention to the scope of their analysis and longevity of their solutions. They must consider not just the intended beneficiaries (e.g., customers, clients, funders) but also non-beneficiaries and what economists bracket out as "externalities" (e.g., distant environments, marginal social groups, future generations, and other species deemed to be beyond designers' responsibility).

Cultural Choices and Alternative Futures

Chapter 8 argues that faith in technological solutions remains strong among entrepreneurs and technological adventurers, and can too often be blinkered. It may, for example, discourage assessment of side effects – both technical and social – and close examination of the political and ethical implications of engineering solutions. Over-confidence in the positive powers of technology may also discourage deeper analysis of social problems and wider-ranging human solutions. Our societal trust in technological problem-solving and techno-fixers consequently demands critical and balanced attention.

This chapter investigates the contemporary momentum of technological faith and assesses its compatibility with sustainable culture in its broadest sense. It documents the discouragingly enduring contemporary faith in fixes, especially when envisaging the future – the perpetually safe haven of technological solutions. Despite the experiences of the past century – promoting, applying, and belatedly recovering from well-intentioned technological fixes – popular faith has remained dismayingly strong. I explore examples of contemporary hubris from a new generation of corporate adventurers, the successors of earlier technological fixers. Their forecasts promise technologies to improve the human body, sustain the planet, or, indeed, escape both entirely. The possibilities are as captivating as were past visions of technological utopias but now evoke more vocal critiques.

Increasingly, as the industrial side effects of pollution and anthropogenic climate change have become recognized in the popular culture of modern countries, environmental sustainability has become a focus of societal concern. As chronicled throughout this book, technological solutions have tended to overlook environmental maintenance, attending instead to more familiar and valued social and economic sustainability. Cultural sustainability, an even less recognized dimension, remains correspondingly less visible in engineering approaches. Critiques of such fixes have identified confidence in technological solutions as evidence for inadequate engineering practice, failures of government policy, and/or unfortunate outcomes of modern consumerism. Adoption of technological fixes has important implications for shared social values, the well-being of wider publics, and the social role of engineers. In short, technological fixes carry negative societal consequences.

Alternative technologies have increasingly been trialled as more suitable choices to ensure sustainability while being adaptive to changing circumstances, in much the way fixes often prioritize short-term and local problems. Along the same lines, economist Ernst Schumacher defined "appropriate technology" as morally responsible innovation that takes account of local social needs, resources, labour, and skills in ways that most technological solutions do not. He argued that popular engineering criteria – efficiency, elegance, and versatility – could work against creating a genuinely sustainable sociotechnical system. Schumacher sometimes referred to his approach as "Buddhist economics," in the sense of incorporating moral and social values into modern systematic problem-solving in much the way as do some Eastern theologies.[47] Diluted forms of this sensitivity are evident in recent initiatives by policy-makers to adopt more ethically overt technologies ("responsible innovation"). This chapter ends by assessing such seemingly socially aware and societally responsible alternatives.

Examining a century of technological change and debate, *Techno-Fixers* argues that present-day societal problems cannot be reduced to mere engineering solutions over the long term: human goals and forms of innovation are diverse and constantly changing.

2 Rational Society: The Technocrats and "Technological Social Design"

Fixes for the Masses

In early 1921, a New York City newspaper reporter interviewed the little-known spokesperson of a young organization. The topic seemed unpromising, and its representative was severe and uncompromising. Howard Scott, chief engineer of the Technical Alliance, spoke about disorganization and inefficiency in the postwar world, and how these problems could be tackled by technical problem-solvers. Rational engineers and technicians, he claimed, would bypass the bankers and politicians to "find out what the American people want, and to get it for them." The reporter had difficulty teasing out the message, evidently because "newspaper men rank in his eyes somewhere along with financiers and diplomats."

But the article recorded the birth of an anecdote of unusual persuasive power, and one that was to be recounted over the following century. It demonstrated succinctly how a technical solution to a problem could leapfrog social, legal, and economic approaches:

> For lack of anything better to say, I asked him a question which every advocate of a new order will recognize as an old acquaintance: "Won't you have to change human nature first?"
> Mr. Scott smiled dryly.
> "Did you have to change human nature," he asked, "in order to keep passengers from standing on car platforms?"

"Go on," I said, "I'm listening."

"They put up signs first," he continued, "prohibiting the dangerous practice, but the passengers still crowded the platform. Then they got ordinances passed, and the platform remained as crowded as before. Policemen, legislators, public service commissions all took a hand but to no effect; then the problem was put up to an engineer.

"The engineers solved it easily. They built cars that didn't have platforms."[1]

As his audience appreciated, the "cars" were streetcars; the "platforms" were the open boarding areas and steps at one or both ends. By enclosing these areas and removing external handholds from which passengers could hang (and fall), engineering design could straightforwardly compel and correct human behaviours. Thus, where legislation and moral exhortations failed, engineers and their inventions could secure desired social outcomes.

The anecdote was timely. Horse-drawn streetcars had been largely replaced by motor vehicles at the beginning of the twentieth century, and passenger safety gained rising attention. American streetcars had begun to incorporate features such as enclosed platforms and pedestrian fenders during the 1890s, and automatic doors and folding steps from the 1910s, although Howard Scott's crediting of beneficent engineers was questionable.[2] When he was interviewed, these improvements were becoming standard features on new streetcars, and older models in major cities were being retrofitted. Production of the mass-produced Birney Safety Car of 1915, in fact, peaked in 1921, and improved designs such as the popular Presidents' Conference Committee (PCC) streetcar were to be introduced as late as 1936.[3]

Scott's first telling of his popular tale contained the seeds of a notion that was to spawn corollaries over subsequent decades. The central message of the anecdote was the superiority of technical innovation over social solutions. The idea entrained confidence in the power of inventions as agents that compel societal change ("technological determinism"). And, as Scott described it, such sensible guidance would also require rational experts at the helm. The reporter asked how he would deal with public opinion:

Mr. Scott let me know that he was vastly bored.

"It is all a technical matter," he said. "It makes not the slightest difference whether the public knows about it or not. The steam engine
didn't need a press agent. The Einstein Theory doesn't require any special legislative enactment. If the only people who can bring order out
of our present industrial chaos find out exactly how to do the job, we
needn't worry about the next step."

This smug confidence in the superior problem-solving abilities of engineers
over other varieties of expert soon became known as "technocracy."[4] Scott's
synthesis envisaged engineers as essential for social improvement and the
advance of modern civilization ("technological progressivism"). Linked to
his potent tale, the hubris of their abstract ideas was contagious.

This single example was to develop a rhetorical life of its own. It was restated, recast, and reapplied to explain the logic of engineering approaches
to new socio-political situations over the following decades, and its content
and style informed the template for later promotion of the technological
fix. Its practised delivery in different venues made the message generic
enough to reduce technocratic ideology to a set of unassailable truths. So
compelling was the tale that it was reproduced in American technocratic
literature into the twenty-first century.

This chapter addresses the prophets and popularization of this modernist
faith. It was largely a story of talk, not action. Its apostles evangelized a growing belief in engineering culture that was to convert ever wider audiences
throughout the century. Tracing the expression and spread of the idea reveals how the first techno-fixers packaged modern technological beliefs for
wider publics.

Engineers of Social Order

Howard Scott was the public voice and chief engineer of both the Technical
Alliance and its successor organization, Technocracy Inc. His compelling
personality and comportment were important in transmitting his ideas (figure 2.1). Scott's rhetorical tone matched his public persona. Anthropologist
Margaret Mead, a close friend throughout Scott's life, described him as "an
extraordinary person, well over six feet in height, gaunt and rangy, Irish and
somehow a man of the frontier, endlessly inventive and *prophetic*."[5] His

Figure 2.1. Howard Scott, chief engineer and seer of Technocracy, c. 1940.

long-time associate, petroleum engineer M. King Hubbert, recalled being "pretty much bowled over with the man's scope, knowledge, understanding" when they first met in 1931.[6] Scott was severe and intimidating. The frustrated *New York World* reporter had complained, "He is an engineer, and he wouldn't argue. He would answer questions if he had the answer, but if he didn't have it, he would express no views."[7] Scott himself gave inconsistent accounts of his professional training, travels, and engineering experience, but investigative reporting revealed rather little of each in his background.[8]

Surrounded by myths, Scott's limited engineering training and experience belied his successful style of engaging with wide audiences. A persuasive and magnetic speaker, his self-confidence, informal speaking style, and fluent command of data on industrial practices impressed his audiences, including established scientists and engineers. Described by historian William Akin as a "bohemian engineer," Scott frequented Greenwich Village in New York just after the First World War and throughout the following decade.[9] There he encountered better known economists, engineers, and scientists who formed the Technical Alliance in 1919. Consisting initially of a group of some seventeen men and women, the loose affiliation included economic philosopher Thorstein Veblen (1857–1929), electrical engineer Charles Steinmetz (1865–1923), conservationist Benton Mackaye (1879–1975), architect Frederick L. Ackerman (1878–1950), and physicist Richard C. Tolman (1881–1948). Most of them identified publicly with what American contemporaries recognized as "progressive" and "reform" policies in the period before and after the First World War. Participants in the Progressive Era sought to replace the self-interested corporate expansion and harsh economic disparities of the Gilded Age with more socially responsible industrial practices. Several of the organization's advocates subsequently were to occupy posts in the Roosevelt administrations during the 1930s and 1940s.[10]

As spokesperson, Scott channelled a current of loosely connected ideas. Veblen, a wide-ranging academic, provided the theoretical core for the group's shared perspective. An early theme of his writing was economic waste and inefficient usage of wealth. Veblen railed against the bad example set by the leisure class's "conspicuous consumption," a term that he coined.[11] Veblen nevertheless argued that modern technologies would ultimately lead to collective social benefits. One of his aphorisms, "Invention is the mother of necessity," can be unpacked as confidence in technological determinism, that is, that innovation creates inevitable social effects. Veblen, in fact, popularized the term in his analysis of the economics of Karl Marx (1818–1883).[12]

Scott's brief anecdote also aligned with broader American intellectual currents of the period. Economist Stuart Chase, arguably the most visible member of the Technical Alliance group in the mid-1920s, published on industrial inefficiency, waste, and consumption.[13] He rebuked traditional production methods and social configurations in factories as ineffective and slow. "Efficiency engineering" was also championed in government throughout the decade, notably by Commerce Secretary Herbert Hoover (1874–1964, and US president 1929–33). Hoover, in turn, channelled the ideas of Frederick W. Taylor (1856–1915), who had promoted the "scientific management" of factory workers and processes.[14] The tale of the safe streetcar fitted the wider narrative of solving old social problems via rational modern design. Confidence in societal progress via wise engineering solutions became a popular feature of American industrial discourse from the early twentieth century.[15]

International Precursors for American Themes

The ideas of American Technocrats were informed by contemporary discussions among European scientists. Key members of the Technical Alliance were immigrants, and they maintained close ties with European peers at a time when European sciences dominated their American equivalents. Veblen was Norwegian-born; Charles Steinmetz of General Electric had emigrated from Germany at a time when socialism was more actively discouraged there. And, despite murky biographical details, Howard Scott also spent part of his youth in Europe, where he likely picked up some of his engineering perspectives.

Contemporary publications were also disseminating international perspectives on sociology, economics, and politics. After the Great War, English chemist Frederick Soddy (1877–1956), discoverer of nuclear isotopes, was calling for an economy based on energy expenditure rather than on monetary flow. He envisaged a technological society founded on fundamental thermodynamic principles and guided by scientific men.[16] The Technocrats were to adopt a near-identical idea in their system of prices founded on "energy credits." Like the pronouncements of the Technocrats, attention to

Soddy's views became more acute after the financial crisis of 1929 and through the subsequent Great Depression. "Born of the troubled times in which we live," he wrote, "there has been growing up from a number of independent and at first quite unconnected roots a group of doctrines broadly described as the application of the system of sciences of the material world, physics and chemistry, to economics and sociology. They have a common feature in that they are all due to the original thought of scientific men." Soddy identified these "more or less independent doctrines" under terms such as "Cartesian, Physical or New Economics, Social Energetics ... and Technocracy," and he offered his own coinage, "ergosophy," as the underlying principle: a "social philosophy arising wholly out of the universally obeyed laws of the physical world."[17]

As he observed, the period after the First World War saw an upsurge of related ideas in European scientific cultures. Soddy's core notion – that social phenomena could be analyzed according to the methods of physical science – nevertheless had older Enlightenment roots. Utilitarianism, for example, was the arithmetic of morality developed by English philosophers Jeremy Bentham (1748–1832) and, two generations later, by John Stuart Mill (1806–1873). Bentham's "Greatest Happiness Principle," or "Principle of Utility," sought to maximize happiness by quantifying positive pleasures and negative harms and weighing them up to guide decision making.[18] The approach was familiar to Victorian engineers in optimizing mechanical designs. Applying rational algorithms to ethics offered speedy and unambiguous means of selecting the best actions.

Nineteenth-century conceptions of sciences applied to the social sphere were shaped by the work of Auguste Comte (1798–1857) in France and the Belgian Adolphe Quetelet (1796–1874). Both had used the term *physique social* ("social physics") to label how the actions of groups of people could be understood. Comte developed a ranking of human knowledge in which mathematics and physics had primacy, owing to their reliance on measurement of quantities. Lower-ranked sciences were chemistry and biology, then dominated by the less analytical techniques of classification and description. Comte characterized two other categories of knowledge – metaphysics and religion, respectively – as inferior because of their imprecision and inability to be refined by rational methods. He argued that all forms of knowledge could be improved by adopting such techniques. Consequently, the new field, referred to as "sociology" by the mid-nineteenth century, adopted the contemporary tools of mathematics and physics. Comte's philosophy of "positivism" valued the discovery of invariant laws of the natural and social world by relying on observation and experiment. Quetelet, in his turn, sought to quantify aspects of human existence to allow for their analysis via the methods of the physical sciences. He promoted the developing field of statistics, for example, by investigating social phenomena such as the near-constancy of the annual rate of suicides in specific locales and the curiously persistent frequency of undeliverable "dead letters" in post offices. Quetelet correlated criminality with social conditions such as age, poverty, and education. His investigation of the measurable statistics of *l'homme moyen* ("the average man") revealed regularities in populations that could be quantified. His approach convinced him that collections of individuals behaved like simpler physical systems according to mathematical laws.[19]

Similarly, "physical economics" grew in Scotland from earlier ambitions to identify physical principles in economic science.[20] Soddy's notions also fitted into scientific discussions of "social energetics" and "social thermodynamics," which identified wider links between human systems and fundamental physical laws that related to energy. "Social mechanics," by contrast, applied the physics of statics and dynamics, then being extended by German physicist Ernst Mach (1838–1916). In pre-Soviet Russia, Alexander Bogdanov (1873–1928) developed "tectology," which more ambitiously sought to combine the biological, social, and physical sciences. Focusing on abstract systems, Bogdanov equated human, societal, and physical laws. Like the nineteenth-century concept of energy itself, this equivalence suggested that people, communities, and their invented world were interrelated and exchangeable. The implication was that technologies could supersede, bypass, or substitute for human capabilities.[21]

The Russian-born American sociologist Pitirim Sorokin described at least four flavours of this "mechanistic school" popular in contemporary sociology during the 1920s. He identified "social energetics," for instance, as a concept developed by Belgian chemist Ernest Solvay (1838–1922) and German physical chemist Wilhelm Ostwald (1853–1932). Sorokin labelled other variants "social physics" and "mathematical sociology." These socio-scientific worldviews identified a close connection between physical laws and descriptions of how society functioned. These were understood as more than mere

Table 2.1
Intellectual currents relating to technology and society

1700s	Enlightenment rationalism: human values and societies can be understood and improved by rational analysis.
1830s	Positivism: scientific knowledge about the natural and social world progresses by acquiring quantifiable evidence. Human nature can be described by empirical measurement.
1840s	Social physics: statistics can reveal mathematical laws of human behaviour.
1860s	Biological evolution: viable forms of life are determined by their physical environments. The varieties of life forms and collective human traits are the consequence of this natural law of selection.
1880s	Eugenics: a society's "genetic health" and evolution can be directed by scientifically selecting desirable traits embodied in chosen breeding partners and preventing offspring from unsuitable partners.
1890s	Taylorism: factory workers can be modelled as components of technological systems. Scientific management optimizes tools and working conditions to shape human actions and maximize output of such sociotechnical systems.
1910s	Technological progressivism: innovations generally produce beneficial social outcomes.
1910s	Social mechanics, social thermodynamics, social energetics, physical economics, ergosophy: an economic system founded on the flow of physical energy is rational and just.
1910s	Technological determinism: innovations irresistibly drive and determine social consequences. Technologies, not human actors, are agents of social change.
1920s	Technocracy: purposeful technological invention can manage society.
1920s	Behaviourism: human behaviours can be understood as being shaped by living environments and life experiences.
1930s	Technological futurism: as depicted by science fiction forecasts, societies of the future will centre on and be empowered by technology.
1940s	Technologies as empowerment: new technologies can transform living conditions to deliver wide social benefits.
1950s	Radical behaviourism: animals and humans can be understood as sophisticated machines, and their behaviours can be shaped (conditioned) by events and environments designed intentionally as learning experiences.
1960s	Technological fixes: deliberate technological solutions are equivalent to, or better than, social approaches, and can substitute for them.
1980s	Technological innovation: in societies designed to accommodate it, technological change can drive the economy and human well-being.
1990s	Technological economics: innovation is the key driver of modern societies.
2000s	Transhumanism: human capabilities can be extended via technologies as a new and directed phase of evolution. The definition of humanity itself can be reshaped by our incorporation of these personal technologies.

analogies or empty metaphors: they identified physical laws as the general bases not only of physics but also of society itself. The proponents thus argued that sociology, as a new science of society, could be founded on the concepts and techniques of the physical sciences. The implications were that conventional approaches would give way to scientifically based analysis and action. Combined with physical economics, the new social sciences aimed to characterize and ultimately manage human affairs.[22]

Among engineers and designers, similar currents were circulating. The Russian Revolution of 1917–19 was evolving rapidly under Vladimir Lenin (1870–1924) and then Joseph Stalin (1878–1953). Artists, engineers, and scientists were reimagining their roles in relation to citizens and the state. Germany, too, was being reconstructed under the postwar Weimar government and was pursuing new initiatives in designing for social utility. The Deutscher Werkbund, for instance, an association of German craftsmen formed in 1907, had sought to establish design principles that combined mass production with efficient function and aesthetic appeal. In architecture, the *Neues Bauen* (New Building) movement, also formed in 1907, promoted the unornamented and functional building style (widely emulated in other countries) later known as modernist or internationalist design. An important example was the Bauhaus school of art, founded by architect Walter Gropius (1883–1969). This fitness for social purpose was an important theme of industrial design and architecture throughout Germany's Weimar Republic from 1919 until Hitler's rise to power in 1933 and had close associations with the Russian Constructivist movement in art and technical design.[23] Bauhaus and related groups proclaimed that the form and function of designs should reflect social needs. These themes illustrate their confidence in the crucial role of invention in shaping human potential and hint at complementary trust in technological measures to improve society.[24]

The summary sketched above, and outlined in table 2.1, suggests the ebb and flow of intellectual currents in scientific culture. While traceable to respected scientists, these notions were neither mainstream nor explicit. Frederick Soddy's list of some of the intellectual threads traceable during the 1920s highlighted their shared convictions not only about scientific laws governing social behaviours but also their lack of mutual coherence.

Technocracy Inc. and the Rhetoric of Rationality

Howard Scott and the members of the Technical Alliance were sampling and synthesizing such ideas emerging among their international peers. Their aim was to apply them as soon as possible in North America. The idea of "pure" versus "applied" sciences was a contemporary cultural theme and, representing the latter, the interests of the engineers were more pragmatic than were the principles proposed by their scientific counterparts. As Scott's first interview revealed, the purpose of the Alliance was "to survey the possibility of applying the achievements of science to societal and industrial affairs." By collecting sound facts and applying rational engineering principles to modern problems, the not-for-profit organization would champion the replacement of "maladministration and chaos imposed upon the industrial mechanism."[25] Founded on the ideas of Veblen, Soddy, and others, the group's message was that technical experts, rather than politicians and financial interests, were the only viable providers of effective solutions for the problems of modern society (figure 2.2).[26]

In the economic and industrial context after the First World War, the ideas championed by the Technical Alliance gained diverse attention in the United States. Labour organizations such as the Railroad Brotherhood and International Workers of the World consulted the group,[27] and the *New York World*, aligned with the national Democratic Party, published its lengthy interview with Scott a year later.

A broader cultural current was the growing valorization of technology in American life. The "Machine Age," a phrase that exploded in popularity during the interwar period, reflected a new pace and confidence for modern society. New inventions provoked expectations of societal transformation.[28] The phrase signalled public awareness of the dependence of urban life on modern technologies and labelled conflicting sentiments about the positive but unavoidable changes delivered by technological change. Scott's rational streetcar was the vehicle by which the inevitable future would be delivered.[29]

Channelled through the persuasive character of Howard Scott, this simple story was to survive organizational shifts. Examples of technological fixes are an important feature of Scott's rhetoric from the earliest communications of the Technical Alliance. While that organization faded from public view through the decade, the financial crisis of 1929 and the deepening eco-

THE *REAL* REVOLUTIONIST

Figure 2.2. Scientists as social disruptors, from *Technocracy Magazine*, 1935.

nomic depression brought its ideas to much larger audiences during 1932–33, a period during which the two major political parties were not taking effective actions. Scott and a handful of former Technical Alliance members coalesced to revive work on an "Energy Survey of North America," intended to analyze national growth in engineering, rather than in economic, terms.

Under the banner of "Technocracy," the cadre worked in vacant rooms at Columbia University's industrial engineering department, where for some eight months they collected statistics on industrial production. By mid-1932, some of their charts and predictions had been leaked to research bodies, labour organizations, economists, and newspapers. Scott again served as the director and charismatic spokesperson for the small cluster of technical experts, and over the next year he was inundated with national attention. As he sloganized it, "the word technocracy, as representative of a new body of thought, means governance by science, *social control through the power of technique*."[30]

The group's dedicated focus and confident explanations appeared to offer a quick route out of the economic crisis. Attracting a strong coterie of engineers alongside wider audiences, it excluded only active politicians and repeatedly disavowed political affiliations from either the left or the right.[31] The organization accreted unaffiliated supporters and interpreters over the following months. Its rapid rise was bolstered by its enigmatic character: as Scott noted in his only national radio broadcast some eight months after the group's formation, "to date, it has written fewer than 14,000 words but, judging from its response on this continent and abroad, those 14,000 words have done their work well."[32] Indeed, the pared-down character of these populist appeals was to become central to the organization's enduring rhetorical style.

Public attention exacerbated divisions within the group, however. Scott's incautious statements and misleading background brought withering criticisms. In 1933 he and a subset of adherents, notably geophysicist Marion King Hubbert (1903–1989), legally incorporated their organization to consolidate their authority. Technocracy Inc. consequently became the official voice of the movement and, more often than not, the personal voice of its "Chief." Scott's communications consequently became the authorized information channel.

Howard Scott's turf was curiously configured. Contributors to the Technical Alliance were American, but Technocracy Inc. had wider membership and cited examples from beyond North America. Technocratic literature described its "continental" perspective. The countries included – in North and Central America, Greenland, and parts of South America – were explained as comprising a self-sustaining geographical region.[33]

From the mid-1930s the newly formalized organization founded local chapters across the United States and Canada. Its compelling sermon was preached via local speeches and lectures, exhibits and membership drives, and regionally circulated newsletters and magazines. Unlike its numerous competitors, Technocracy Inc. was a survivor. Dominated by the views of Howard Scott until his death in 1970 and beyond, its "continental headquarters" (CHQ) shifted successively from New York State to Pennsylvania to Georgia and, during the 1990s, to a small town in the state of Washington nearer the centre of mass of its remaining supporters in the western United States and Canada.[34] With its hierarchical direction but reliance on grassroots activities, throughout the century the organization provided a remarkably stable and sparsely authoritative message. This simple administrative configuration and carefully controlled content contributed to the wide dissemination of some of its ideas and its survival into the internet era.

The most enduring of the organization's notions was faith in technological solutions for complex social, political, and economic problems, and the compact expression of what others came to call "the technological fix."

Life History of a Twentieth-Century Parable

Scott's 1921 anecdote about streetcar design may have been a regular feature of his private conversations before the emergence of the Technical Alliance, and certainly of public addresses after the incorporation of Technocracy Inc. With the example of streetcars replaced by railway carriages, for instance, it is recorded some sixteen years after the *New York World* interview in another of Scott's speeches:

> People say you can't change human nature. We of the engineering profession approach it in another way. The only method of regulating has been to prohibit. You have noticed the sign, "Passengers are prohibited from standing on platforms," in railway cars. Engineers came along and designed a train without platforms and said, "Stand on them if you can."

Issues of coercion and control melt away, he suggested, when replaced by benign physical environments that ensure safety. The same lecture gave a second example of engineering design that prudently guided appropriate social behaviours in factories. Instead of signs prohibiting dangerous use of equipment, he showed the picture of an accident-proof press. "You cannot be hurt by any operation of the machines. Put your hand in, and it won't work. Even cigarette smoke will stop it. The product can be made responsible."[35]

In effect the machine, rather than the operator, has agency. It embodies moral authority rather like a parent constraining the behaviour of a wilfully disobedient child. Yet Scott never analyzed his rhetorical anecdotes further and seldom multiplied them. The canonical example of public transport appears again in a 1952 speech, but now linked to the fashionable topic in American psychology of behavioural conditioning:

> You see in the matter of conditioning, remember the old railway coaches that had the metal sign on them? They're still running around. "Passengers are prohibited from standing on the platform of this coach. It's contrary to law." Well, that's your legalistic, moral approach. It forbids people. That isn't the scientific, technological approach. You design a car without a platform and say stand on it if you can. Very simple.[36]

And again, during the late 1950s, the example reiterates the efficacy of engineering over morality and politics. The casual oratory now hints more overtly, and perhaps smugly, at the intellectual hierarchy of wise designers versus an obstinate public:

> You see the technological approach to these problems is totally different than the moralistic, arid, legalistic approach … Well, the engineer just designs the car without a platform, and he says stand on it if you can. You're a sucker if you try it – it isn't there.[37]

Scott's delivery in each of these cases differs from his more general argumentative ploy of displaying graphs and quoting industrial statistics that indicate seemingly inevitable trends – in particular, the impending failure of capitalist economics.[38] Curiously, the positivist reliance on quantifiable evidence is replaced by an almost religious faith. The concise sketches are akin to New Testament parables: vaguely situated allegories that were seemingly universal in their applicability to new situations. The structure and aesthetics of the narratives and anecdotes arguably accentuated the appeal of these ideas. Delivered verbally and graphically with the imprecision of everyday language, they suggested common-sense truths having an archetypal generality. Like the best parables, Scott's tales provided revelatory insights that appeared, in retrospect, self-evident to his audiences.

The rhetorical form of the parable traditionally compares and contrasts, with "bad" versus "good," in this case, being exemplified by wilful human misbehaviours versus astute engineering, respectively. Like traditional parables, which communicate a moral or spiritual message, Scott's tales express modern realities with an overarching judgment: engineering designs effectively compel social change and circumvent resistance, and consequently should be recognized as the most beneficent means of ensuring societal improvement. The stories contrast ineffective and wrong-minded societal actions – prohibiting, regulating, and mandating – with the automatic social controls imposed by rational designs and their sage (and morally responsible) designers.

The parable-like role of the anecdotes and evangelistic tone of the public meetings is also suggested by Scott's deportment as an impressively tall, deep-voiced, and revered figure always addressed by acolytes as "the Chief" and by one local chapter's collection of such writings and speeches after his death into a publication of biblical import, *The Words and Wisdom of Howard Scott*.[39] His secular sermonizing was typically extended by opportunities for collective enlightenment, where deferential audience queries received lengthy and discursive responses from the Chief. The sessions disseminated technocratic ideology in an appealing demotic style usually supported by technological aids. Indeed, throughout the 1960s, most of Scott's interactions with members of the organization were in the form of long-distance telephone question-and-preaching sessions that followed some local chapter meetings.

The streetcar anecdote appears to have originated and remained with Scott himself. It is notably absent, for example, from the uncredited *Technocracy Study Course* written by the organization's co-founder M. King Hubbert (figure 2.3).[40] There, a relatively pale alternative is recounted instead to communicate the potency of technological determinism:

It is seldom appreciated to what extent … technological factors determine the activities of human beings … Thousands of people cross the Hudson River daily at 125th Street, and almost no-one crosses the

Figure 2.3 Marion King Hubbert, Technocrat, petroleum geophysicist, and economic forecaster, c. 1935.

river at 116th Street. There is no law … It merely happens that there is a ferry at the former place which operates continuously, and none at the latter. It is possible to get across the river at 116th Street, but under the existing technological controls the great majority of the members of the human species find the passageway at 125th Street the more convenient. This gives us a clue to the most fundamental social control technique that exists.[41]

What Scott had exemplified as clear-headed public protection ensured by thoughtful design becomes an anodyne technique of social regulation. In fact, Scott's paternalism and commonsensical tales contrast with a colder, *Brave New World* tone in Hubbert's *Technocratic Study Course*.[42] One of its lessons ("20. The Nature of the Human Animal") discusses at some length Ivan Pavlov's behavioural experiments alongside contemporary findings in endocrinology, arguing for the rational conditioning and shaping of social behaviours by the methods of science. By implication, popular beliefs and actions are shaped predominantly but chaotically by their social environments and must give way to the rational methods of experts able to engineer those environments to achieve desired behaviours. The scientific tone and appeal to rationality were central elements of the message.[43]

Common-Sense Truths and Iconic Imagery

Sometime during the 1930s, a graphic was prepared by Technocracy Inc. to illustrate the original version of Scott's anecdote.[44] According to the present-day administrator of the organization, the streetcar image "along with a dozen others depicting mankind's evolution into the technological age was displayed along the upper walls on every Section Headquarters on the continent." It featured thereafter as one of the paradigmatic illustrations used to explain the key idea of technocracy in public outreach programs (figure 2.4).[45]

Versions of the illustration were also reproduced in post–Second World War Technocracy publications and carried various captions. Scott's example had been multiplied to serve distinct rhetorical purposes, often stretched to fit new symbolic functions. The parable featured in a 1945 article on streetcar usage in Vancouver, Canada, for example, to illustrate the more general claim of the inevitable progress of technologies and their consequent positive societal benefits. What safe streetcars had contributed to modern orderly cities, it suggested, other technologies would assuredly multiply.[46] In 1946, Technocracy member Leslie Bounds used the streetcar as a looser analogy to illustrate a general technique for eradicating crime. The caption was:

> We can end chiseling and the greater proportion of crime in America, not by passing laws or the greater efforts to enforce existing laws, but by the simple expedient of making it impossible and unnecessary to commit the crimes.[47]

Technological innovation would supplant laws and bypass traditional behavioural techniques such as moral guidance, education, and prosecution. Alongside this bold prediction, the author repeated a more widely accepted forecast. Urban regeneration would replace slum neighbourhoods, a technological transformation that would alter living contexts and, he argued, inevitably circumvent the human behaviours that resulted from them.

Member Walter Palm reused the streetcar graphic in the *Technocrat* magazine two years later to argue more broadly and allegorically that

> it is futile to attempt to solve the social problems of this continent by business and political methods. An entirely new design is needed. Since our problems are technological, only a technological solution is adequate. The chart … illustrates the simplicity and ease with which problems are solved by our scientists, technologists and engineers.[48]

And, in 1952, the *Northwest Technocrat* employed the same illustration to accompany an article on dangerous practices of transporting livestock to market:

> This Chart depicts graphically Technocracy's scientific approach to our social problem; a technological, physical solution for what is fundamentally a purely physical problem. Result, greater safety and comfort, and the elimination of "crime."[49]

Cattle, groups of people, and society at large, it showed, could be safeguarded by shrewd technological guidance. Nevertheless, in an era when

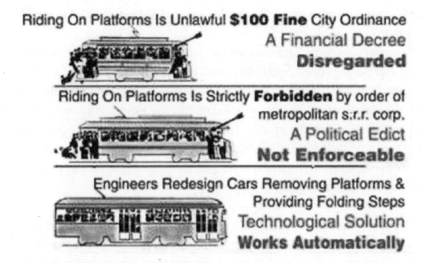

Riding On Platforms Is Unlawful **$100 Fine** City Ordinance
A Financial Decree
Disregarded

Riding On Platforms Is Strictly **Forbidden** by order of
metropolitan s.r.r. corp.
A Political Edict
Not Enforceable

Engineers Redesign Cars Removing Platforms &
Providing Folding Steps
Technological Solution
Works Automatically

Figure 2.4 Technological solutions to social problems: rational streetcar design according to the Technocrats, c. 1935. The origins of the four-colour streetcar graphic, reprinted in post-card size, remain unclear.

tramways were being buried under asphalt to dedicate urban streets to buses and cars, the streetcar example no longer illustrated the leading edge of engineering wisdom and beneficence but, rather, the past. Even so, while the potency of the example faded, Technocracy Inc. did not update its message. The course materials, public exhibits, and lectures remained based on imagery conceived between the wars.[50]

Showcasing Technocracy

Prior histories of the Technocrats focus on their brief popularity and aberrant characteristics. I argue instead that their potent idea of technological solutions for society remained influential throughout the postwar period and went mainstream.

As suggested by the organization's iconic example of rational streetcars, the Technocrats had remained attentive to the single authoritative voice of Howard Scott. Members' postwar analyses mapped new situations onto the archetypes of the interwar period. The simple narrative structure and chain of command made the organization's message simultaneously doctrinaire and weakly relevant to the postwar world. But the organization disseminated its message of "technological social design" via new routes to entice wider publics.

To attract audiences, Technocracy Inc. employed a communicative technique that Scott dubbed "symbolization." First mooted in 1937 and trialled in mass demonstrations two years later, it was systematized in the decade after the war. Centred on iconography, motorcade processions, and visual and audio spectacles, the practice had arisen among members emulating Scott's own public presentation, but it was rapidly theorized by key participants. Symbolization sought to arouse curiosity and to attract fresh audiences by subliminal appeals:

> The balance of the public is not interested and is incapable of assimilating the necessary facts and implications ... At the proper moment, the trained organization of Technocracy may find it necessary to present Technocracy to the masses in assimilatible [sic] form.[51]

To present-day observers, the techniques suggest both interwar political rallies and modern advertising techniques. Symbolization, according to Scott, was more effective than the commercial advertising of the day; it was designed holistically to condense the organization's themes and ideology into rapidly absorbed visual representations. Uniform dress, machine-like social organization, and the monad (a yin-yang symbol in vermilion and chromium) became visual metaphors for the logic and modernity of technological solutions to societal problems. Employing searchlights, road signs, billboards, radio broadcasts, and exhibition trucks, the processions represented the regimentation and efficiency of the organization. For contemporaries, the members' adoption of regulation grey suits and vehicles evoked comparisons with European fascism, but the Technocrats stressed the role of modern technologies and media to suggest their matter-of-fact and emotion-free scientific rationale. Like the streetcar graphic, the varied methods of symbolization were designed to win over audiences by sidestepping analysis.[52]

The organization had been banned in Canada as a divisive political force in 1940 when the British Empire, but not the United States, was at war, but the sanction was lifted three years later. For similar reasons, M. King Hubbert's appointment as a senior analyst for the American government's Board of Economic Warfare in 1942 was challenged by a civil service commission enquiry in 1943, which ruled him unsuitable for the post owing to his links with Technocracy Inc.[53]

In the summer of 1947, Technocracy Inc. implemented Operation Columbia, an all-out "symbolization activity" that was to be repeated across North America over the following decade. Seeking to demonstrate the renewed postwar vitality of the organization across national borders, the symbolization involved a fleet of several hundred regulation-grey vehicles making a two-week membership drive, round-trip lecture tour, and cavalcade through cities along the west coast between Los Angeles and Vancouver, Canada.[54] It was supported by mass mailing campaigns in the cities to be visited, an associated "Tech Net" of shortwave radio amateurs to maintain communications along the parading convoy, some one-hundred "sound cars" carrying public address systems, and truck-mounted war-surplus searchlights to attract local audiences. Grey-suited volunteers distributed technocracy literature at the roadside to passing cars and visitors on foot and in rented arenas in major cities; Vancouver alone drew some five thousand paying attendees.[55] At each venue, an exhibit bus displayed iconic posters, including the ubiquitous streetcar graphic. The following summer, Operation Golden Gate attracted some twenty-five hundred Technocrats from around the United States to parade around the Bay Area in four hundred cars and buses, followed by a road tour to Los Angeles. Subsequent publicity spectacles on a less ambitious scale included Chapter-organized motorcades and public picnics through the early 1950s.[56] Either first- or second-hand, the local activities and press attention surrounding symbolization events likely provided the first exposure of many young postwar North American designers and planners to technocratic ideas. The power of technologies to transform social life and governance, the leitmotif of the technocratic literature, meshed with postwar thinking.

From Safer Streetcars to a Logical Society

So, between the turn of the twentieth century and the end of the Second World War, ideas and rhetoric championing technological approaches to societal problems became simpler. These amorphous confidences had emerged from scientific and technical workers shaping the new culture of modernity, but their simplification can be attributed to a persuasive prophet, Howard Scott, and his organizational vehicles, the Technical Alliance and Technocracy Inc. Reduced to compact articles of faith, the ideas could be readily expressed, promoted, and absorbed.

Scott's claims about the role of technology and technologists in society can be summarized as follows:

1   Social problems in modern society are caused, and ultimately solved, by technological change.
2   Conventional tools – notably economics, politics, and education – are relatively ineffective.
3   Rational technological change of environments can produce new social behaviours rapidly.
4   Only technically competent people are equipped to solve modern social problems.

Expressed as a modern truism, Scott's streetcar anecdote encapsulated these ideas. When he first related it, the example cited a contemporary transformation familiar to citizens in modern cities. The rational tale of cool engineering logic triumphing over undisciplined publics impressed those very audiences. And the message embodied in the simple iconography could be readily absorbed, so much so that details of the increasingly nostalgic streetcar designs were scarcely noticed over the following decades. The success of the lesson in wider culture owed much to its brevity and superficial generality, making it impervious to contemporary analysis and historical revisionism.

Of equal note is how effectively the notions of "technological social design" (later to become the strapline of Technocracy Inc.) were disseminated. The simple messages were refined to their essentials by years of repetition. Aiding their spread and survival was the lack of documentary precision or elaboration, allowing them to be translated into allegories fitting new contexts and arguments. Communicated matter-of-factly by Howard Scott as a handful of parables, the simple anecdotes were immune to criticism and easily recounted or displayed by supporters. He rehearsed his examples without extension or correction through his final years.

Scott's rhetorical claims were potent and fertile. The idea that technological design could be the most effective and rapid means of solving societal problems seemed to Technocrats and many engineers of the period to be self-evident. This confidence entrained even more influential hidden assumptions, for example, that the inevitable benefits of technological solutions are generally beyond dispute and that technologies necessarily determine social outcomes. Faith in progress and the doctrine of technological determinism were implicit but readily accepted implications of Scott's anecdotes. Thus, the style as much as the meagre content of the messages delivered wide-ranging beliefs to non-engineering audiences as well as to technological experts.

Their promotional campaigns illustrate the careful attention Technocrats paid to their style of dissemination to persuade broad audiences. The identification of parables, icons, and proselytizing is helpful for understanding how brevity, imprecision, and imagery proved not only compelling but also inspirational for audiences over decades. The episodes suggest the value of close reading of such popular discourses to better understand their role in the growth of modern cultural dogmas. Scott's promotion of technological fixes is salutary in illustrating how the power of simply expressed ideas communicated by confident technical experts can shape the beliefs and actions of generations.

3 Mid-Century Confidences

From Technocratic Parables to Inventive Communities

What do interwar "sound bites" have to do with preaching technological faith throughout subsequent decades? Howard Scott's streetcar anecdote and illustration not only swayed a small coterie of adherents but also circulated widely among American wartime and postwar technical workers. The message of Technocracy Inc. was spread via local chapters, periodicals, parades, and public meetings. It had particular appeal and publicity in southern California, where Richard L. Meier worked as a chemist during the Second World War. And, from 1933, a Technocracy Chapter was active in Chicago, the locus of the Manhattan Project group in which Alvin Weinberg worked during the war. Both individuals crafted more mainstream versions of the message for new audiences.[1]

Thanks to the symbolization drives, the visibility of the Technocrats actually grew after the war. It was a time when young American professionals such as Meier and Weinberg were rapidly adapting to an unfamiliar environment of national labs and government-sponsored research projects (a working domain that Weinberg later called "Big Science"). The fresh setting encouraged larger-scale thinking and the application of scientific knowledge to new social contexts. In 1945, both Meier and Weinberg were early members of the Federation of Atomic Scientists, a group of Manhattan Project scientists and technologists who championed the need to apply science more effectively to societal improvement. And Weinberg's subsequent career at

Oak Ridge, Tennessee, had close ties to the Midwest and industrialized northeast, which, along with the western United States and Canada, remained the heartland of Technocracy Inc.[2]

This chapter traces the consolidation of these once radical confidences in the wider culture of orthodox science and policy. It focuses on three themes: the wartime ascendency of technological methods of problem-solving; its application to the postwar reconstruction of more livable cities; and the boldest and most articulate voice for technological solutions, scientist and social planner Richard Meier. Describing himself as a technological optimist, Meier's scholarly approach to technological solutions for societal ills was the antithesis of the promotional techniques of American Technocrats. In common with many of his generation, wartime experiences and postwar opportunities shaped his confidence in technological fixes.

Wartime Lessons: The Value of Technical Elites

The recent war had developed two themes originally championed by the pre-war Technocrats, but on an international scale. First, new technologies could solve urgent problems and, second, technologists were essential to this.

Military ascendency had been traded back and forth by the opponents over the six years of the war, providing a surfeit of strategic problems for which technological innovation provided quick, if temporary, solutions. Aircraft were improved from fast to faster single-wing fighters, higher-capacity bombers, and (late in the war, on the German side) jet-engine V1 and rocket-borne V2 weapons. Radar and radar countermeasures developed rapidly. And founded on the highest science of all, nuclear physics, the United Kingdom, the United States, and Canada, outpacing Germany and Japan, developed two varieties of atomic bomb that ended the war. More than anything else, this frenetic technical progress convinced governments and their citizens about the power of inventions to shape political outcomes.[3]

The waging of the war increasingly resembled the Technocrats' notion of "technological social design." Wars traditionally had depended on the organization of human combatants, and wartime leaders now acknowledged their heavy reliance on technological innovation and scientific guidance to refine strategies. In Britain, Winston Churchill relied on a scientific adviser, physicist Frederick Lindemann (Lord Cherwell, 1886–1957), over some

twenty years straddling the Second World War and had almost daily contact
with him during the last half of the war. Lindemann headed a wartime
statistics department, S-Branch, which summarized complex questions in
numerical and graphical form for the prime minister. These mathematical
methods aimed to translate and codify findings to allow quick decision making. Lindemann's faith in scientific methods and quantitative evaluation of
human systems, and his approach, which involved reducing multivariate
problems to single actions, owed much to industrial management predecessors such as Frederick W. Taylor and his disciple Henry Gantt.[4]

More generally, British scientists and engineers during the war applied
mathematical techniques to support military decision making. Besides Lindemann, key members included physicist Patrick Blackett (1897–1974),
chemist Henry Tizard (1885–1959), zoologist Solly Zuckerman (1904–1993),
and Robert Watson-Watt (1892–1973), developer of radio detection and
ranging (radar). Their subject, later known as "operational research" (OR),
applied statistics, mathematical modelling, and logical inference to problems
such as bombing tactics, camouflage design, and anti-tank weapons. Even
more secret were British efforts to break German codes. While code breaking
during the First World War had relied on experts having backgrounds in
languages and the humanities, the work in the Second World War relied on
mathematicians such as Alan Turing (1912–1954) as puzzle-solvers and on
electrical engineers of the General Post Office, long responsible for the national telegraph and telephone networks and radio licensing, who devised
some of the earliest digital computers to speed up processing.[5]

Significantly, these scientific methods also assessed human dimensions
of complex systems, such as considering observers and operators as components of radar detection and communications networks. Code breaking,
too, was reliant on human procedural errors to reveal the weaknesses in the
designs of electromechanical encrypting machines. By characterizing
human operations as machine-like, technological solutions were rendered
more reasonable. The upshot of problem-solving on the part of wartime
physical scientists was that military problems were addressed by routinely
reducing them to systems analysis. Sociotechnical systems and their human
components, it appeared, could be competently handled by physical scientists and engineers.[6]

In the United States, Franklin Roosevelt came to rely on equivalent scientific advisors during the war, who advocated similar solutions for wartime ascendency. His closest associate was Vannevar Bush (1890–1974), professor of engineering at the Massachusetts Institute of Technology (MIT). Bush chaired the National Advisory Committee on Aeronautics (NACA) in 1938 and lobbied for the creation of the new National Defense Research Committee (NDRC) in 1940, a year before the United States entered the war. His NDRC sought to marshal civilian scientists to achieve military goals. Bush appointed James Conant (1893–1978), president of nearby Harvard University, as a member of the committee. When a more powerful coordinating body, the Office of Scientific Research and Development (OSRD), was formed the following year, Bush became its director, and the NDRC was subsumed into it with Conant as its chair. Bush, supported by Conant, reported directly to the president and effectively managed national resources for wartime research and development, most notably steering the Manhattan Project.[7] They drew upon their peers in the fertile academic environment of Cambridge, Massachusetts, to spearhead wartime projects. The MIT "Rad Lab," developing applications of radar, was the largest such wartime research laboratory. After the war, the region became the first major locus for government-funded technical research in the national interest. What had begun as engineering quick fixes for tactical problems were reconfigured as a generic national and corporate tool. In 1951, the opening of Route 128 provided a home ground for innovative engineering companies fertilized by postwar government contracts. As discussed in chapter 4, MIT was to remain a reservoir for expertise in government defence consulting after the war and, arguably, provided the seeds for confidence in technological fixes for American policy.[8]

For such scientists and their government masters, wartime innovation carried an evident political lesson: with the clarity of urgent priorities, technological experts and their solutions could streamline a wider range of complex problems. This required engineering expertise to be available at the top of organizational hierarchies. But it also seemed to rely on other societal adjustments familiar to wartime publics: a simplification of problems to their technical dimensions, decision making from above, and a subdued democracy.

Postwar Science and Technology for National Progress

For those who played a role in governing and managing the war – both vic-
tors and vanquished, and across the political spectrum – the postwar period
saw a growing consensus about the social value of technological innovation.

The public revelation of the atomic bomb and the wider promise of
atomic energy captivated legislators, citizens, and technologists alike. The
Manhattan Project demonstrated that high science and scientists could be
harnessed to societal goals speedily and effectively. The three collaborating
countries – the United Kingdom, the United States, and Canada – had delivered two distinct forms of a powerful weapon in just three wartime years,
dramatically extending scientific knowledge and engineering expertise. Dedicated scientists, it showed, could meld scientific curiosity with problem-
solving in the national interest. The mystique of these elite experts was
amplified by the enduring secrecy surrounding atomic energy during the
early Cold War, which kept the engineering details under wraps.[9] For the
British public, a more accessible version of the myth was that of the "boffin,"
portrayed as a technical expert who persevered in developing valuable technologies – radar, penicillin, bouncing bombs, and the like – despite social
ineptness and limited resources.[10] Echoing the ideas of the Technocrats, such
experts were both essential and atypical citizens.

For policy-makers, the most important lesson was that wartime technologies illustrated the value of government-funded technical research to
achieving social goals. Beginning in 1946, new national government-funded
laboratories grew from the temporary wartime facilities in the United States,
Britain, and Canada, all dedicated to accelerating research on atomic energy.
The Manhattan Project became the postwar model of large-scale, multi-
centre research projects, inspiring not only the Kennedy administration's
space initiative of the 1960s but also the Nixon administration's War on
Cancer of the 1970s.[11]

Technological innovation became synonymous with longer-term problem-solving and societal progress. Atomic energy was touted as offering lim-
itless energy to supply ambitious societal aims. Cozy Arctic cities might be
created, new canals excavated, food supplies multiplied, and transportation
revolutionized.[12] But the achievement of creating an atomic bomb exemplified a particular case of an expanding role for science and technology in society. Two other wartime successes, penicillin and radar, were just as rapidly

being generalized into a suite of technologies having civilian benefits. New antibiotics were already saving lives, and radar was making air travel safer and defending national sovereignty.[13]

Among the most enthusiastic promoters were those who had been part of the Manhattan Project. Scientists and engineers had worked at Argonne Laboratory near Chicago, designing the first nuclear reactors; at Oak Ridge, Tennessee, exploring the properties of the first prototype reactors and developing methods of amassing radioactive uranium and plutonium on an industrial scale; at Hanford, Washington, tending the mammoth production reactors; and at Los Alamos, New Mexico, where atomic bombs based on uranium and plutonium fission were designed and tested.

These wartime locales had been makeshift, uncomfortable, and unsustainable, but the implications of the work were revolutionary. The unfamiliar working environment of secrecy and urgency encouraged the participants to voice their views about the uses made of their technologies and to seek new roles as advisers to government. At some sites, new grassroots organizations represented not only the collective technical expertise but also the developing social and policy views of scientists and engineers. Examples included the Association of Oak Ridge Engineers and Scientists, the Atomic Scientists of Chicago, the Association of Los Alamos Scientists, and, in the United Kingdom, the Atomic Scientists' Association. At the war's end, the most senior of them were feted as scientific heroes, and their public exposure provided a platform for them to influence not just science but also visions of a scientifically led society. Some of the members and groups within this "atomic scientists' movement" coalesced into the Federation of Atomic Scientists two months after the atomic bombs were dropped on Japan. They were united not only by concerns about the threat of atomic weapons to destroy human civilization but also by an eagerness to explore the societal applications of peaceful atomic energy. The atomic scientists argued for a role in cooperative international policy-making and the opportunity to foster human progress at home through innovations such as irradiated foods (to prevent spoilage) and new harbours created by nuclear explosions. The spirit of internationalism and progressive ideals linked high science with social benefits and promised a brighter future for all.[14]

The end of the war thus saw a rare conjunction of interests for government, technologists, and the wider public. An important factor in creating this receptive atmosphere for science as a tool of government and society

Figure 3.1 Technology for human progress at the museum of the atomic city, Los Alamos, 1954. Jungk, *Tomorrow Is Already Here*, fig. 6. During the late 1980s, Jungk (1913–1994), an Austrian journalist who explored the implications of early atomic energy, was also known for the "Future Workshop," a technique for grassroots development of social solutions in complex systems.

was a report by Vannevar Bush in the closing months of the war. President Roosevelt had asked him to provide advice looking forward, based on his administration of the Manhattan Project and the Office of Scientific Research and Development. Roosevelt asked specifically how wartime scientific knowledge could be made public "for the improvement of national well-being"; how scientific medicine could be further progressed; how government could aid scientific research; and how scientific talent could be fostered for peacetime benefits. The result was an iconic document, "Science: The Endless Frontier," which shaped the attitudes of the public, administration, and engineering industries over the following decades. In it, Bush concluded that science should be a central concern of government and that government-supported research was necessary to improve public health, housing, and agriculture. "A stream of new scientific knowledge," he argued, was needed "to turn the wheels of private and public enterprise." The postwar United States would be driven by technical experts applying scientific methods to achieve government-defined public goals.[15] Bush summarized a new consensus, subtly mutating ideals championed by the Technocrats some twenty years earlier.

This merging of science with human affairs – a prerequisite for later acceptance of technological fixes – can be traced throughout postwar thinking. For instance, the link between wise governance and science first explored by various European scientists at the turn of the century re-emerged at this time from the new field of "cybernetics," which was developed during the postwar years to explore general principles of systems of control and communication. Like the notion of technological fixes, the underlying ideas combined ancient approaches and empirical techniques with modern rational perspectives. It also related biological and engineering perspectives to bring together the analysis of animals and machines and, more broadly, the understandings of control mechanisms in the animate and inanimate worlds.[16]

The term, from the ancient Greek κυβερνητική, meaning "governance" or "steering," referred originally to social systems and specifically to self-governance. But the word also labelled persons in control of hardware, such as the helmsman of a ship who steers a steady course despite changing winds and currents. The idea was readily extended to denote a non-sentient mechanism that could maintain equilibrium or control. A well known example is the centrifugal governor, invented by James Watt in 1788 to control the speed of steam engines. A rotating shaft with two attached and suspended weighted balls is connected to the engine output shaft. If the engine speed increases, the weighted balls swing out away from the shaft and, via a lever arrangement, reduce the flow of steam into the engine to lower its speed. More recent examples are thermostats to regulate room temperature and the combination of sensors, motors, and electronic controls that enable drones to hover.

Such clever devices are forms of technological fix. They replace the human tasks of observation, decision making, and regulation via an automatic contrivance. This substitution of human control with a technological innovation is a powerful example of how cybernetics transcended disciplines to reimagine technical, biological, and even social systems. Generic engineering ideas introduced from the late 1940s, such as the concepts of feedback loops and servomechanisms, entered popular language and altered perceptions of the role of technologies in translating human roles into machine equivalents. In more abstract terms, cybernetics identified mathematical and physical relationships in complex systems. The field of study applied these insights to achieve specific goals, which were as diverse as automatically controlling the level of water in a toilet tank or preventing circumstances in which mob violence might occur. The new principles, generalizing what had been clever particular solutions, were mirrored by the popularization of technological fixes a decade later. This conflation of physical principles with human behaviours encouraged trust in technological solutions to social problems.[17]

Redesigning Cities to Deliver Well-Being

Technological quick-fixes were also timely. The Second World War had created urgent human problems on a vast scale. The destruction of European and Asian cities during the war was followed by a quarter-century of rebuilding and, in some places, wholesale redesign. This rational reconstruction prioritized urban housing and infrastructure as the most effective means of improving the well-being of citizens. To varying degrees, urban redesign acknowledged (and later challenged) the efficacy of technological fixes for solving social problems.

Designers reconstructed and arguably improved ravaged cities in distinctive and locally specific ways. In Japan, reconstruction was complicated by desperate food and housing shortages. Systemic factors – fiscal restraints and an uneasy sharing of power between municipal and central governments – were further hampered by occupation by the US Army. Nevertheless, the Tokyo Metropolitan Government ambitiously planned (even if it did not accomplish) to redesign the city beyond the devastated districts and to reconfigure one-third of the urban area as greenbelt. The rebuilding of Osaka revealed similar tensions between municipal planners and the central government. The claimed social benefits of technological changes, including wider roads and redevelopment of port facilities, were disputed by local residents who were alienated by both levels of government planning. The special case of Hiroshima, almost completely destroyed by the first atomic bomb, revealed concerted action by planners and citizens but a similar lack of agreement. Reconstructing postwar living environments evidently did not so much improve lives as merely sustain them.[18] In short, this technological fix overlooked wider human issues.

Cost was a limiting factor, too, for European regional governments impoverished by the war. Surviving water, gas, electrical, and sewage systems were more often repaired than redesigned for future benefit. Some towns responded to pressures from residents to regain their prewar land rights, living spaces, and lifestyles. The heritage of old towns and traditions encouraged other cities to more gradually reconstruct districts and facsimiles of lost landmark buildings. Yet other urban centres seriously damaged by the war considered ambitious replanning that aimed to improve quality of life more fundamentally.[19]

The scale of proposed improvements was most evident in Germany. This can be attributed both to the scale of wartime destruction, epitomized by the Allied bombing campaign that levelled Dresden, and to optimistic prewar modernist themes circulating among German designers. Adolf Hitler had envisaged grand redesigns of German cities from 1940. His personal architect, Albert Speer (1905–1981), set up a working group in 1943 to prepare for postwar urban reconstruction. Their plans for triumphalist architecture, monuments, and parade grounds had little to do, however, with direct social benefits. At best, the schemes sought to reinforce deeper cultural attitudes: a focus for national pride, identity, and state power at the heart of the party's appeal. Earlier German movements, as discussed in chapter 2, had predated the Nazi urban visions by promoting more socially directed design.

In Britain, the war came to be associated with correcting social injustices at home and abroad. In 1943, the London County Council (LCC) prepared a short film, *The Proud City*, which described how the living conditions of greater London had been increasingly questioned as German bombers had

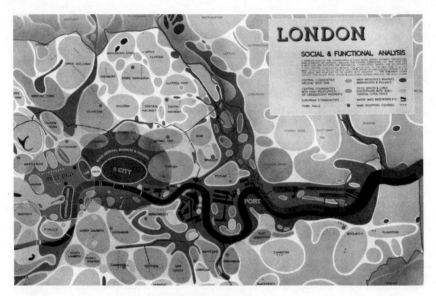

Figure 3.2. Social progress through rational design. Social and functional plan (detail) for a revamped postwar London, 1943.

progressively devastated the city. Voice-overs from members of the public asked whether the prewar social problems in the city were locked in by physical conditions and whether those conditions could be improved by coherent urban design. "Must we always have slums, and an East End and a West End?"; "Three people killed in the streets every day"; "Parks and gardens in some parts, but in other parts, only the streets for us kids to play in." The film argued for rational central design of the material environment to improve community life. It publicized design work begun by the LCC in 1941, the second year of the war, to prepare a grand plan "that would bring a new order and dignity" to the urban region. It described social research – about inadequate living, working, and transport conditions; health and leisure – that would be ameliorated by technological change. Scenes of a drawing office depicted how "our architects ... work on their designs for a new London."

The redesign, directed by J.H. Forshaw (1895–1973), architect for the LCC, and town planner Patrick Abercrombie (1879–1957) of University College London, was published that year (figure 3.2). As Abercrombie observed: "London grew up without any plan or order; that's why there are all those bad and ugly things that we hope to do away with." The plan envisaged compact high-rise flats, open spaces and facilities for leisure, rationalized road routes, and rezoned industry – in short, "order and design" to replace "muddle and confusion."[20]

Similar confidence in the power of rational design inspired Abercrombie's contemporaries. In 1945, Robert Bruce, chief engineer of the Glasgow Corporation (city government), called for the demolition of tenement housing in the Scottish city's Victorian centre to begin again from scratch. Town planner Patrick Abercrombie responded to the Bruce Plan a year later with his own contrasting but equally bold proposal of redistributing half the city's population to outlying new towns.[21] Abercrombie was also adviser for other regional and city authorities, including the war-ravaged east coast city of Hull. There, he proposed widening and straightening the river, relocating the fishing industry seven miles nearer the North Sea, and rationalizing the road network and shopping districts. Like the other postwar urban plans, the reconstructions of London, Glasgow, and Hull proved expensive and unpopular for some stakeholders and were delayed and scaled down to be carried out piecemeal.[22]

The plans emphasized common-sense observations about how urban conditions could restrict and limit human pleasures and aspirations, but they also hinted at tenets of faith for the designers. The flip side – that reimagined environments would generate social improvements – was much less obvious. At the heart of this assertion lay unmeasurable uncertainties. How much space, and how little commuting time, was required to create thriving neighbourhoods? Would wider roads and community centres, for example, trigger enduring social improvement? Or would populations adapt to create new muddles and ugly things?

A second article of faith – that planners could rationalize the disordered material world and, with it, the social world – was implicit but even more questionable. Reliance on central planning carried political associations that were simultaneously appealing and distasteful. The management of the world wars illustrated the effectiveness of such planning but also raised the hackles of populations required to bend to higher authorities to meet national goals. Abercrombie himself, in fact, had been of two minds about wise managers versus democratic desires. He had suggested after the First World War that civic associations could work alongside the familiar local authorities. These citizens' groups would bolster localism, "forward[ing] the progress of decentralization which is the foundation of local advancement ... the necessary corrective during the reconstruction period, of the excessive centralization that has been brought about by the war."[23] Such conundrums were among the first signs of popular resistance to the developing notion of the technological fix.

Richard Meier and the Science of World Development

The complex problems of postwar reconstruction, uniting legislators, economists, and planners to redesign ravaged cities, inspired larger-scale technological corrections. A key figure in this ambitious thinking was Richard L. Meier. Later described by Alvin Weinberg as a "professor of everything," Meier was in fact a wartime research chemist who turned to investigating technological solutions to the problems of social development (figure 3.3).[24] He was a founder and later executive secretary of the Federation of Atomic Scientists (later renamed the Federation of American Scientists) in 1945, which attracted Manhattan Project technologists and scientists at Los Alamos and at Oak Ridge, including Alvin Weinberg. The organization aimed to assert the authority of technologists in modern governance: to educate and influence policy-makers and to encourage the peaceful application of science and, specifically, nuclear energy for national and international benefit. Meier and Weinberg consequently shared prominent influential acquaintances such as physicist Leo Szilard but moved in separate career circles.[25]

Meier established a career as a wide-ranging academic. He described his investigations as being triggered by a 1948 publisher's grant to appraise "the potential new solutions to the world problems of nutrition and comfort." Other grants came from business sponsors, including the Pabst Brewing Company and the Ford Foundation, for research in the behavioural sciences. Developing his ideas during a Fulbright Fellowship in the postwar UK, Meier conceived technological means of tackling problems associated with poverty, social organization, and city infrastructure. Meier taught in the University of Chicago Program of Education and Research in Planning from 1950 to 1956 and was a researcher in systems theory at the University of Michigan

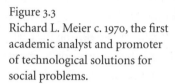

Figure 3.3
Richard L. Meier c. 1970, the first
academic analyst and promoter
of technological solutions for
social problems.

until 1967. Thereafter, at the University of California at Berkeley, he applied
systems analysis to inform principles of urban planning and regional development – a refinement of the more ad hoc technological fixes pursued
by postwar city designers.[26]

Like Scott before him and Weinberg later, Meier was a technological optimist who was oriented explicitly towards planning technological systems
to reduce inequity and to yield wider societal benefits. His publications were
detailed empirical studies supporting extended arguments. In his first book,
published in 1956, Meier explored technology as a means of providing "re-
alistic utopias … consonant with the resources at the disposal of society."[27]
He sought to apply postwar discoveries in science and technology to create
"technically sound procedures" for world development. His preface cites
atomic power and the mass culture of algae as contemporary examples of
technologies "groomed for a leading role in the improvement of human welfare" by supplying bountiful energy and food to growing populations.[28]

Sustaining human needs: *New Patterns of Living*

In countries still mired in postwar reconstruction, Meier's focus on a science of world development was uncommon. The United Nations, formed in October 1945, had established the Economic and Social Council (ECOSOC) as one of its original six bodies, but its focus was policy review, recommendations, and coordination of economic and social development. There was initially little direct involvement of scientists and technologists as advisors or key personnel. Instead, such experts more often operated during the early Cold War period in the service of their countries: either in new government laboratories, freshly funded university departments, or large firms. Their products were nationally oriented: science and technology for security, research and development for national renown, consumer satisfaction, and corporate competitiveness.

Meier himself noted that international perspectives were equally compartmentalized. He criticized "fragmentary and disorganized" research by scientists and engineers that did not fit together "the implications of new data in order to establish their various possible effects upon society." The prospects for world development, he noted, had been viewed "from the limited standpoint of the geologist-geographer, soil scientist-agronomist, international economist, etc.," leading to incompatible conclusions.[29] The disconnection among academic disciplines meant that "the concepts of the working natural scientist do not mix well with those from the social scientist," and he argued for an interdisciplinary approach:

> The invasion of social studies was carefully considered; it was intended to establish a continuity of analysis that overlaps the experience of social scientists. Only by such means (i.e. the penetration of each other's domain) can the various specialists communicate with each other in order to solve problems of society.[30]

By crossing academic boundaries, Meier proselytized, the scientific fraternity could create broad models to direct policy. He proposed analysis that would go "one or two stages farther than ha[d] been possible before" by focusing on *socio*technical connections: how technological innovation could contribute to human solutions, and how social and economic institutions could adjust to technological powers. Devoid of rhetoric, his sensitive research program was an academic successor to the brasher claim championed by the Technocrats – that technological innovation was the most effective basis for solving societal problems. But between the lines was a progressive version of the popular notion of technological determinism: technologies would produce inevitable social effects that could be tailored for the benefit of the weakest members of society.

Meier's sermon was ambitious and universal: to apply scientific knowledge and engineering expertise to identify solutions to universal human problems. "Many of the requirements can be stated in scientific and technical terms," he suggested. "Scientists and engineers can consider the various ways in which the nutritional targets, or shelter targets, might be reached [and] they can look for short cuts."[31]

His attention to the essentials of life was well timed. Wartime deprivations in Europe and Asia had been witnessed or directly experienced by millions of citizens and combatants. They were still freshly remembered by, or ongoing among, wider home populations in many countries. Given their recent experiences, more affluent audiences could empathize with the endemic poverty of inhabitants of the less developed world. Meier set himself to this task with zeal. He analyzed average human nutritional requirements by age and dietary composition. The need for clothing materials, for instance, determined largely by standards of decency and convenience, turned out to be between two and thirteen kilograms per person per year, depending on regional climate and housing conditions. Consumption of fuel, energy, construction materials, and metals was similarly estimated.

Each statistic inspired technological improvements. Human water consumption, for instance, depended on climate, topography, and soils. Growing populations could be supported by systems of wells and pumps, watersheds, reservoirs, and – thinking to near-future technologies – desalination plants to generate potable water from seawater. Similarly, Meier evaluated food consumption as satisfying physiological needs but as being constrained by cultural practices. He cited margarine as a then unpopular example of an artificial replacement for butter. He noted that more successful changes in dietary patterns tended to be associated with the intermixing of communities and dominated by the stronger culture. This observation seems to have informed his approach: *new* forms of food might gain favour because of their *lack* of associations.

This attention to social factors consequently led Meier to new technical solutions. "Our best strategy," he reasoned, "is to go back and take a closer look at new technology." The logical inferences appeared compelling. Members of impoverished communities could not afford adequate amounts of conventional protein sources; the result was physical lassitude, which prevented self-improvement. This vicious cycle could nevertheless be broken by "incorporating certain fruits and vegetables into a new, balanced dietary pattern" or by supplying inexpensive synthetic vitamins and enriched foods.[32] Yet the analysis demonstrated the gulf between clever technological fixes and social and cultural traditions. Meier was intrigued to discover, for instance, that plant leaves could supply vast protein content. He cited the then current development of Multi-Purpose Food (MPF) by the California Institute of Technology, based on soybean meal as the starting material.[33] Meier devoted continuing attention to other forms of food production – harvesting deep-sea shrimp, industrially processing plankton, and adopting large-scale hydroponic farming methods – that would reduce costs but probably introduce "food prejudice." Such solutions demanded social adaptation to technological changes that were literally unpalatable:

> There must be an institution that would persuade the farmer that it was in his own interest to alter his activities ... [but] it must be admitted that the prospects of rapid improvements of supply are rather poor ... When the whole social system on the rural scene must be revamped in order to achieve a doubling or trebling of the yield protein, then it cannot be expected that such increases will be achieved in less than a generation.[34]

The tenor of Meier's argument was that such social resistance could be reduced by designers' sensitivity to traditions but that, in any case, established practices would inevitably succumb to the rational improvements. This was a softer form of technological determinism than that promoted by some of his contemporaries: an empathetic acknowledgment of human values and frailties, combined with the firm conviction that the medicine was necessary.

Fuel supply was also a prominent candidate for Meier's rational analysis. Indeed, energy had been the scientific core of social thinking by turn-of-the-century physical scientists and the Technocrats, as explored in chapter 2, and was to be the career focus of Alvin Weinberg in the second half of the century. Meier enthusiastically explored forms of energy production and sought to determine options for satisfying growing societal demand. But, again unlike his contemporaries, he was eager to explore wider sociotechnical adjustments. Just as artificial foods should not seek to merely mirror conventional food sources, he reasoned, so fuel demand could be traded off for other adaptations. For example, agricultural production of either foodstuffs, or cotton for garments, or biofuels might be optimized. More insulated or better ventilated buildings might reduce energy demand. Wearing more or less clothing could achieve the same end. In cold climates, warmer clothes reduced dietary demands, too: Meier calculated that "every calorie devoted to [producing] fibers is returned at least three- to six-fold in the saving of food." He lined up the engineering aspects of clothing ("thermal comfort, freedom from irritation, launderability, cost") against cultural requirements such as convenience (especially pockets) and identification of social status.[35]

Meier's bigger picture of technological systems offered more scope for designers than did contemporary engineering disciplines. But with scale and abstraction came a consequential inattention to human qualities and social outcomes. In places, his single-mindedness was overt: "physiologists and psychologists," he observed, "are now coming to view a human being as a nerve network with a body wrapped around it so as to stabilize the system, giving it opportunity to develop its organization of impressions."[36] This was an extreme form of reductionism, which simplified humans to biological machines (and was largely compatible with the contemporary themes of both behavioural psychology and cybernetics). Much of his first book, in fact, discusses human needs as one might plan a biology lab. While white mice in such circumstances are supplied with the quantifiable essentials of life, their social requirements are typically neglected. In the same way, Meier suggested that his analysis of essentials unleashed human potential: "providing comfort permits the human organism to seek individual and social ends at full efficiency." Yet this provision could be further optimized. Comfort, he suggested without irony, "may involve the use of drugs or illusions rather than important quantities of energy." Thus, the organism could accommodate to less costly technological solutions: "The use of drugs to control motion sickness, for example, appears to be a much simpler and more adequate solution than incorporating gyrostabilizers into all transport equipment."[37]

Humans are malleable, Meier suggested. "As the various societies in the world become informed about comfort and the most convenient means of achieving it," cultural traditions would likely be discarded. He praised, for instance, the rapid pursuit of new technologies ("plastics, fibers, biochemicals, pharmaceuticals, and refined metallurgy") that were transforming postwar Japan.[38] In short, technologies could provide greater happiness for the greater number. Meier's ideas echoed Jeremy Bentham's utilitarianism, which had confidently proposed trade-offs to maximize happiness.[39]

As Meier admitted, his focus on comfort and convenience was a materialistic view of human behaviour. The logic of design options dominated his analysis, diverting his attention away from the unquantifiable adjustments required of individuals and their society. As these examples suggest, he reasoned that a top-down approach was essential. Climate scientists, agriculturalists, architects, chemists, and physicists were his favoured sources of information. From their data, technical experts would rationally determine optimal engineering solutions, and populations would adapt to them. In its goals and methods, Meier's confidences shared much with those of the Taylorists, Technocrats, and behaviourists.

Technical Answers for Social Questions

These currents directed Richard Meier's analysis towards new waters. Midway through his first book, he addressed social problems directly. He described turning social dilemmas into scientific and engineering problems, which would be more easily and quickly solved by technological means than by redesigning social systems and reshaping cultural norms.

Meier's enthusiasm periodically broke through his sober analysis. He suggested that new two-way communication technology could faithfully trans-

Figure 3.4 Satirizing forecasts of technology-enhanced learning.
Villemard postcard series, 1910.

mit messages to large populations; this centrally managed process would replace unreliable word of mouth and garbled messages. And "communications technicians," he reasoned, could "bring forward solutions" to the problem of education in a rapidly changing society. They would identify "economical means for transcribing the lesson into widely understood symbols, and then use communication devices ... for conveying these symbols."[40] The example drew upon a familiar trope of his time that remains appealing today: that new learning technologies would guarantee better education. The notion, in retrospect recognizable as a classic technological fix, had been a recurring theme for decades: promoting the use of magic lantern lectures for Sunday schools in the late Victorian era, film projectors in school classrooms of the 1920s, training of army recruits by stereoscope slides in the 1930s, and radio programs to replace chalkboard lessons in the 1940s.

In the decades after Meier's book, classroom television, computers, and internet access were successively touted as technological solutions to the persistent social complexities of education. This confidence spawned the acronym TELT (Technology-Enhanced Learning and Teaching), a currently growing field in educational studies.[41] An example is the One Laptop Per

Child (OLPC) initiative founded in 2005 by Nicolas Negroponte (1943–) of the MIT Media Lab. The program aimed to develop a one-hundred-dollar laptop computer to provide a potent technological fix, revolutionizing education and opportunities in the developing world. Despite the high expectations of the participating companies, the initiative failed to meet its objectives. It was criticized for exporting American solutions to dissimilar African and South American economic, political, and social contexts, for example, and for its reliance on a model of design and distribution that disempowered developing countries.[42] It would be interesting to know what Richard Meier would have made of more recent developments and their unanticipated side effects. Within scarcely a decade of the rise of internet-based social media – briefly lauded as a new democratic route to citizenship – peer-to-peer sharing provided new routes for sharing misinformation during the 2016 American presidential election. The phenomenon might have underscored for Meier his conviction that centrally managed information flow has societal advantages, even if it might reduce the social pleasure of unrestricted opinion sharing.

Meier's breathtaking technological confidence is also demonstrated in his musing about what was later called "carrying capacity." He estimated that the earth could support some 50 billion persons if societies adopted microbiological food production and accepted "nearly complete urbanization over the flat portions of the world." This stark modernist scenario is uncomfortably suggestive of science-fiction dystopias such as *Soylent Green*, in which growing populations of the early twenty-first century are sustained by synthetic food.[43]

Meier addressed the "Malthusian dilemma" (population rise until limited by scarcity of resources) by considering numerous technological options for birth control. He acknowledged cultural dimensions such as the correlation between family size, illiteracy, and male authority, but he argued that a technological solution ("a small pill that produced temporary sterility") would sidestep these factors. Such a pharmaceutical was then under development, being first approved in the United States in 1960 and shortly thereafter in other Western countries. In a later book, Meier noted that these new pharmaceuticals had medical side effects and were not always effective but argued that the scientific approach, rather than social measures, was the only effective alternative to growing populations. Nevertheless, usage remained highest among relatively affluent populations. A correlation between the availability of oral contraceptives and world population has been difficult to discern.[44]

Meier's range of technical attention was remarkable. He concerned himself with the optimal design of household furniture alongside conceptual plans for regional industrial complexes linked to mammoth atomic power plants, while speculating on how populations would adapt to these new necessities. In the decade when television was first being taken up in American homes, he imagined how an electronically integrated system of communications, entertainment, and banking, identifiably similar to the modern internet, would transform society. The process, he argued, was unstoppable and required continuous planning to ensure public adaptation. "Readers will have experienced increasing disquietude [about] a preoccupation with the technical solutions," he noted, but "this has been quite necessary." Actioned by new technologies, rapid development would allow a society "to escape the limitations of its own culture and tradition." Even so, "very soon many new difficulties appear. These, too, can be solved, but the keys to the solution are drawn from experience *outside* the culture, from the accumulated body of scientific and technical knowledge possessed by the world at large."[45]

At the heart of Richard Meier's first book was the conviction that human populations and their needs would continue to grow and that scientific analysis and technological innovation provided the only effective tools to sustain them. Between the lines, his message was that a technological society would usurp cultural traditions and empower a shared vision of global human needs and potential. The process would become deterministic, producing unavoidable changes and the need for vigilance in seeking technological solutions to emerging problems. This notion of embarking on a course that could no longer be avoided or abandoned echoes ideas that were then becoming familiar to French audiences. The work of Jacques Ellul (1912–1984) had a harder tone. It warned that twentieth-century technology had become autonomous, identifying it as the key driver of society, which could not be resisted and perhaps not even controlled.[46]

Meier's enthusiasm divided audiences. As early as 1951, his optimistic technological determinism influenced an economist colleague to praise Meier's predictions as being "so clearly in prospect as to seem inevitable."

He confidently forecast that such rational planning "of course will correspond with a decline in nationalism and the gradual disappearance of economic and political inequalities."[47] An astronomer-reviewer focused on the science, describing Meier's first book as cogent analysis "that placed the various parts of the analysis in perspective," but he chided Meier for not promising guaranteed solutions from fusion power, then being touted as the forthcoming successor to nuclear fission: "there will be sufficient energy," he assured, "for all purposes for everybody in the future."[48] Even some sociologists had praise for Meier's work. W.F. Cottrell contrasted the "naïve belief in the efficacy of constitutions and pacts" that had existed after the First World War with Meier's alternative, showing the world to be "a communications net-work and ... a technological system." "Social technologists," summarized Cottrell, would *bypass* political manoeuvring "without cataclysm or great conflict." His review illustrated the soft technological determinism and technocratic overtones then popular even among scholars of the social sciences: "If scientific planners could induce the earth's people to do that which technology requires for the most efficient exploitation of its resources they could provide a satisfactory way of life." As Cottrell implied, Meier's approach had all the elements later to be described as a technological fix. His principal criticism was that Meier, although "no naïve utopian," did not explain how "those with power ... may be induced to change their ways." Cottrell did not see technological adoptions as inevitable. He concluded that Meier's greatest contribution was to show the price to be paid "to secure the technologist's dream."[49]

British economist Alec Cairncross (1911–1998) more diffidently described the book's "muted iconoclasm" and the "naked vision of social problems in engineering terms" but saw it as reminiscent of the "superb confidence and imagination" of H.G. Wells.[50] For others in the postwar world, Meier's writings evoked both forward-looking possibilities and retrograde ideology. They were, for example, castigated by Marxist American economist Paul Baran (1909–1964) as "reflecting naïve rationalism or the spirit of technocratic speculation."[51] Despite the attention gained by Meier's writing, more of his contemporaries probably favoured another reviewer's opinion of his second book: "factual in his reporting, but oftentimes his wordy evaluation of the scientist in his culture outweighs the task he has set himself."[52] His audiences were a world apart from the Technocrats: they were academic economists, planners, and international aid agencies. While making little impact on popular audiences, Meier's carefully detailed examples were to be reinvented and echoed a decade later in a much streamlined form in the writings and speeches of Alvin Weinberg.

4 Engineering Culture: Alvin Weinberg and the Packaging of Technological Fixes

Fixes for Governance

Technological faith spread from populist fringe groups and closeted academics to American elites during the 1960s. The postwar decades established new alliances between technologists, government, and corporations, a configuration of vested interests identified disparagingly as the "military-industrial complex" by outgoing president Dwight D. Eisenhower in 1961.[1] The most ardent spokesperson for this influential audience was Alvin Weinberg, longtime director of Oak Ridge National Laboratory. The shared working environments of Weinberg and his contemporaries had included interwar progressive science, wartime military technology, postwar National Labs, and subsequent science policy advising. These evolving contexts shaped how these key actors and interpreters envisaged the societal benefits of science.

Weinberg had to work hard to convince his audiences (and himself) that the notion of the technological fix was sound and valuable as a general approach to problem-solving. His promotion of techno-fixes was positioned between, and eventually combined with, two other neologisms that generated enduring interest. During the early 1960s, he adopted the term "Big Science" to frame his analysis of government-funded, nationally oriented science. And from the end of the decade, his term "trans-science" labelled problems that required social as well as scientific methods.

Weinberg's promotion and merging of these ideas made technological fixes alternately compelling and contentious for varied audiences. Reflecting the rhetoric of interwar and postwar Technocrats, he packaged a common-place notion and promoted it as the centrepiece of revitalized Big Science. His actions thereby reconfigured the overt politics and social aims of inter-war scientists into a strategy of science policy. Weinberg's exploration of these ideas has continued to evoke discussion and has shaped the canon of science and technology studies and syllabuses of history of technology over the half-century since they were coined.[2]

Alvin Weinberg as Voice of Postwar Science

Born in 1915, Weinberg obtained his first two degrees in physics from the University of Chicago during the 1930s and subsequently taught for two years at a junior college in the city. His career interests were broad, as sug-gested by doctoral studies in the emerging field of mathematical biology under Nicolas Rashevsky (1899–1972). His goal of modelling periodicities in cell metabolism was ambitious in relating the tools of physics to the living world.[3] Through his supervisor, Weinberg developed interests in mathemat-ical sociology and, via Rashevsky's acquaintance, Alfred Korzybski (1879–1950), pursued studies of semantics and logic. These pursuits supported his later activities as public speaker and essayist.

Weinberg gained his substantive working experience as a physicist during the early years of the Second World War, being directly involved in designing the first nuclear reactors for the Manhattan Project at Oak Ridge, Tennessee, and Hanford, Washington. His mentor and life-long friend Eugene Wigner (1902–1995) later recalled Weinberg as "a natural diplomat." Wigner identi-fied these wider competences and orientations as Weinberg's key strengths: "We set him to work on broader scientific questions. His grasp of human personality won over many doubters. He never failed us."[4]

Weinberg's links with Rashevsky and Korzybski were later to channel his first enunciation of the technological fix, but these unconventional contacts attracted the attention of postwar security officers. His interview provides a glimpse of Weinberg's "generally liberal" politics, later to be subsumed within his views on technology. He recalled his political philosophy being "flavored by the spirit that pervaded the college campus," contributing to organizations "in which there were certainly Communists … but after join-ing I somehow felt my conscience had been salved." Security officials also scrutinized Weinberg's wartime associations with groups "dissatisfied with the conduct of the [Manhattan] project." He responded that this "group of younger scientists" was concerned with "the impact of what atomic scientists were doing in the world at large." In summary, he noted, "I have been in the past, if you like, a person on the Progressive side, although, unfortunately, I seem to be getting more conservative as I grow older."[5]

While Weinberg's social concerns mirrored those of many of his contemporaries, his postwar activities were directed towards disciplinary and career ambitions. Oak Ridge became a national laboratory (Oak Ridge National Laboratory [ORNL]) in 1946, but, as technical director from 1948, Weinberg had to lobby actively for the facility's continued work on the design and applications of nuclear reactors in competition with the Argonne National Laboratory near Chicago.[6] He co-organized what was known informally as the "Clinch College of Nuclear Knowledge" (1946) and attracted sponsorship from influential students such as Commander Hyman Rickover (1900–1986), who was to foster the nuclear navy. The subsequent Institute of Nuclear Studies (ORINS, from 1947) and School of Reactor Technology (ORSORT, from 1950) predated unclassified university teaching of the subject by a decade (figure 4.1).[7]

While operating within a strict security environment, both ORNL and ORINS were outward-looking. Their staff, students, and visitors represented American companies, government services, and academic affiliations. ORINS was directed by academics from eight American universities spanning not just physics and chemistry but also biology and medicine. Its lab facilities, travelling speakers, and graduate training program disseminated atomic expertise to the first generation of postwar technologists, managers, government contractors, and military staff.[8]

Assuming overall direction of ORNL in 1955, Alvin Weinberg's management style further expanded his professional networks. He implemented an annual review system for the burgeoning departments, ranging from chemical technology to radiation biology, reliant on external assessors and all-Lab "information meetings." His expanding contacts led to reciprocal arrangements, including roles in external reviewing of academic departments and membership on advisory committees of organizations such as the National Academy of Sciences (NAS). By mid-career, Weinberg's committee work and hand-picked review panels provided a rich network of con-

Figure 4.1 Alvin Weinberg, champion of the technological fix, teaching at the Oak Ridge School of Reactor Technology, 1954.

tacts. As his status grew, Weinberg sponsored those he admired for membership in the Washington, DC, Cosmos Club, which promoted science, literature, and art for the American intellectual elite. He consequently was able to survey not only the nascent field of nuclear science and engineering but also the changing terrain of government-funded science via the perspectives of other disciplines. And Weinberg's growing predilection for public speaking provided a platform for his evolving views about the integration of technological innovation with the goals of modern American society.[9]

Touchstones for Weinberg's Ideas

Both in his personal papers and autobiographical publications, Alvin Weinberg acknowledged his reliance on peers to assess and refine his notions, a strategy that also underlay his "information meeting" approach to Lab management. His ideas typically were shaped by a sequence of personal insights, one-to-one soundings with colleagues, and trial speeches to diverse audiences. The role of mentors and confidantes was crucial to these iterations. While these informal advisors shifted and expanded throughout his career, only a handful contributed to what was to become the "technological fix."

The earliest and most enduring of these advisors was physicist Eugene Wigner (1902–1995). Wigner had been Weinberg's immediate superior during the Manhattan Project, responsible for designing the first pilot plant nuclear reactor at Oak Ridge, Tennessee, and the mammoth plutonium production reactors at Hanford, Washington. After the war, Wigner had served as research director of ORNL, a post passed on to Weinberg two years later. A political conservative, Wigner railed against lab bureaucracy and positioned himself outside the mainstream of American government policy by lobbying for civil defence measures alongside nuclear weapons.[10]

Weinberg's public adoption of his mentor's views was illustrated by his serving on the President's Science Advisory Committee (PSAC) on Civil Defense during the Eisenhower and Kennedy administrations, and implementing a civil defence research project at ORNL in 1964 with Wigner at the helm. That project is notable in being the only example in Weinberg's career in which social measures were to be vaunted over technological approaches. Aiming to look at civil defence "from the broadest possible viewpoint," it would consist initially of "about a dozen mature natural and behavioral scientists ... to develop a coherent picture of the whole Civil Defense problem." The project led to the employment of sociologists at ORNL and the opportunity to directly inter-compare social science and engineering approaches to solving societal problems.[11]

Harvey Brooks (1915–2004) was Weinberg's second most influential guide during the 1960s. Dean of engineering and applied physics at Harvard, Brooks had received his education and experience during the Second World War. An exact contemporary of Weinberg, he had participated in designing postwar nuclear reactors and developed an interest in applying scientific expertise for policy-makers. Weinberg and Brooks first interacted via an ad hoc committee on reactor policies and programs for the Atomic Energy Commission in 1958, on PSAC from 1960, and later through the National Academy of Sciences' Committee on Science and Public Policy (COSPUP), which Brooks chaired.[12] As his obituary put it, Brooks was interested "in the sociopolitical context of science," seeking "to understand both the human and the more intellectual dimensions of a rational society," with science being "driven by societal needs." Weinberg was to credit Brooks with clarifying his ideas about technological fixes as a positive rational route for society.[13]

Two other scholars had a role in nurturing Alvin Weinberg's ideas about science and society. The first was Yale historian and bibliometrist of science Derek de Solla Price (1922–1983). Weinberg first encountered de Solla Price's writings in 1961 in the midst of developing his critique of "Big Science," and he recommended him as a forecaster to a presidential advisor.[14] In committee work and his first public exposure via speeches and newspaper articles, Weinberg had been reflecting on the implications of government-funded science and, specifically, the challenges of adequately funding the explosively expanding variety of worthy projects. He had already courted controversy by arguing that the prioritization of Big Science was having a demonstrably negative social impact, notably questioning whether the space program was more valuable than medical research to discover more effective treatments for cancer.[15]

De Solla Price's historical extrapolation of scientific activities appeared to support Weinberg's conviction that the trajectory of science had to change, and quickly. He struck up a correspondence with the historian, with topics ranging from the resource constraints limiting scientific expansion to the deterioration of scientific writing style, musings that were to inform Weinberg's subsequent speeches, essays, and committee work over the following five years. Weinberg discovered that his own views, founded on a career in engineering and administration, accorded closely with those of the humanities scholar. The two organized a six-week summer institute at ORNL on "Humanistic Discussions in Science" in 1963, teaching alongside Manhattan Project physicist Robert Oppenheimer (1904–1967), anthropologist Margaret Mead (1901–1978), and historian of technology Melvin Kranzberg (1917–1995). Weinberg's writings on Big Science were mirrored by de Solla Price's book quantifying its historical dimensions (figure 4.2). Buttressed by this scholarly back-up, his first public foray into science policy yielded favourable reviews from the press and encouraged further provocative reflections on science and society. Indeed, he was to confide that "writing controversial essays" about Big Science was "a risky business" and that he was "reconciled to reading violent reviews from my colleagues who differ fundamentally with my viewpoint."[16]

Weinberg's other significant adviser was Emmanuel Mesthene (1931–1990), director of the Harvard University Program on Technology and Society, which had been set up with IBM funding in 1964. Mesthene himself was an academic philosopher and former RAND Corporation researcher and, like Weinberg, confident of the role of technology in shaping the future both in terms of improved social structures and more effective political initiatives. Harvey Brooks was a contributor to his Harvard program, and Mesthene and Weinberg were members of Brooks's NAS Committee on Science and Public Policy. Mesthene was a sounding-board for Weinberg's private musings on technological fixes from 1965 and was later to publish the first book anthology that included Weinberg's initial writing on the subject.[17]

But Weinberg found similar sentiments among his peers. His ideas echoed and amplified voices in the Eisenhower, Kennedy, and Johnson administrations, which increasingly identified social problems as national targets and consulted closely with scientific and engineering experts. Those administrations were just as closely attuned to the achievements of the Manhattan Project. The waging of war in Vietnam arguably had technocratic overtones in its converging focus on measurable parameters and hardware, and growing reliance on technological development and technical specialists. For example, Robert McNamara (1916–2009, US defense secretary in the Kennedy and Johnson administrations) had been a participant in Statistical Control for the Pentagon during the Second World War, where he identified budgeting inefficiencies that limited the progress of the war. The activity was a refinement of a much older practice: military logistics, more generally termed supply-chain management. By broadening its mathematical remit, McNamara's office adopted the methods of Frederick Lindemann's more ambitious S-Branch in Britain. A key simplification for both was to describe human portions of complex systems as aspects of engineering. McNamara viewed progress in Vietnam via his familiar economist's lens, quantifying progress in terms of body counts and territory gained.[18]

McNamara's management of the Vietnam conflict was consequently receptive to scientific experts. Among the most influential were the elite scientists making up the government advisory group known as JASON. Its

Figure 4.2
Weinberg on the challenges of "Big Science," a term that he popularized.

remit was to independently brainstorm scientifically informed solutions to sensitive and high-security issues. In 1966, the Jasons proposed to counter guerilla warfare with an electronic net of sensors to detect enemy movements, an idea consequently known as the "McNamara Line." A second cluster of scientific experts, the Cambridge Discussion Group based at Harvard University and MIT, recommended technologies that included chain link fences, chemical defoliants, night vision devices, and "area-denial weapons" such as mines. McNamara himself asked for barriers "that wouldn't require additional troops to construct or guard," resulting in the Jasons' recommendation of an "Air-Supported Anti-Infiltration Barrier," consisting of noise makers dropped from aircraft and acoustic detectors with radio transmitters parachuted into the trees along the Ho Chi Minh Trail. When the noise makers – small bomblets designed to injure but not kill – were triggered by the

Figure 4.3 Technological approaches to government policy. Meeting of President's
Science Advisory Committee (PSAC), 1960. Behind President Eisenhower at left
centre is Alvin Weinberg (1), Harvey Brooks (2), Glenn Seaborg (3), Jerome Wiesner
(4), later president of the Kennedy administration's PSAC, and George Kistiakowsky
(5). I.I. Rabi (6) is seated to the far right. Kistiakowsky and Wiesner were members
of the Cambridge Discussion Group, a cluster of physicists at MIT and Harvard;
Rabi was a member of JASON, a government advisory group set up by Eisenhower
to focus on sensitive high-security problems. Both groups were later to recommend
technological approaches for waging the Vietnam War.

passage of troops or vehicles, the human sounds picked up by microphones
would be analyzed by computer to alert an immediate air strike of cluster
bombs.[19] Although based in California and Massachusetts, respectively, and
representing distinct political allegiances, the Jasons and Cambridge Discussion Group were not mutually exclusive: members of both also contributed to the President's Science Advisory Committee (PSAC, Figure 4.3).[20]
Such actors and episodes suggest the dominance of technological perspectives in the close-knit American scientific and policy circles of the period,
their confidence in leapfrogging conventional strategies, and the fertile
ground in which Alvin Weinberg nurtured his own ideas.[21]

For legislators and 1968 presidential candidates, Weinberg eventually proposed a wider national strategy founded on technological fixes. He argued

that the expertise in physical science and engineering marshalled at national labs since the war could be reoriented to address predominantly social problems.[22] With national oversight, he suggested, technological analysis and problem-solving could trump conventional social, political, economic, educational, and moral approaches.

Gestating the Technological Fix

The nascent idea developed in an environment that provided attention and considerable admiration for Weinberg's views. His public addresses during the early 1960s focused on two favourite topics. First was his teasing out of the implications of Big Science. As he reiterated, the new scale of research would demand prioritized government funding, but "technological gigantism" would also require scientists to be more socially aware of, and responsible for, the consequences of applications and to consider the potential for accidents.[23]

Weinberg's second hobbyhorse was scientific literacy, and it ranged from criticisms of jargonistic technical writing to calls for better science education in schools. He was also becoming more conscious of the power of language and presentation. He counselled an acquaintance in the Office of Science and Technology on coining catch phrases, advising that a memorable slogan could suggest the linkage of ideas even if it did not reflect reality. Through these avocational ventures, he attracted positive mentions from scholars in the humanities and social sciences as well as praise from newspaper editorialists.[24]

From the early 1960s, his views on Big Science provided Alvin Weinberg with a platform for greater influence. His authority as director of a national lab allowed a degree of outspokenness and favoured his expanded public speaking and press coverage; work in advisory committees encouraged his musings on the role of science in society; and interaction with a spectrum of scientists and scholars drew him towards what he later summarized as interest in the "the philosophy of science and the government of science."[25]

Further networking accentuated Weinberg's confidence. He spent February 1966 on sabbatical at the University of California at Santa Barbara to prepare a collection of his essays later published as *Reflections on Big Science*. There, he wrote, "I became more and more involved with various segments of the faculty – in particular, the Philosophy, Engineering, Biology, and

Physics Departments ... a sort of roving critic who was willing to express ideas ... the University got its money's worth out of my stay." And there he encountered other forthright scholars, notably philosopher Garrett Hardin (1915–2003), later author of provocative environmental essays.[26]

Specific events further stimulated Weinberg to focus his ideas about technological solutions for societal problems. He had attended the Second Pugwash Conference on Science and World Affairs in 1958 and not only renewed contact with free thinkers such as Leo Szilard but also made new acquaintances with Soviet counterparts such as Alexander Topchiev, both of whom were founder-organizers. The experience may have inspired Weinberg's first publicly expressed views linking technology to social consequences: a 1960 newspaper interview about the products of the national labs. Military technologies, he suggested, could ensure political stability. This commonplace idea – widely shared if seldom stated compactly – was successively sharpened as Weinberg began to identify more examples.[27]

The earliest and arguably most contentious such case concerned the Watts riots in August 1965. Harsh police treatment of an African American motorist had led to community outrage in the Los Angeles neighbourhood and a week of public unrest, arson, and looting that the California National Guard quelled at great cost in lives and property.[28]

Two days after the riots had ended, Alvin Weinberg wrote to an influential contact, Donald Hornig (1920–2013), chemist and special assistant to President Johnson for science and technology. In thinking about the riots, he said: "I began to wonder whether there was any cheap engineering fix that was likely to reduce the probability of their occurrence at least for a sufficient time so that more profound social changes could have a chance to improve the situation permanently." He suggested that "the clue to the engineering fix comes from the observation that the riots seem to come at the hottest, most uncomfortable time of the year," and he recommended using antipoverty funds or money from a projected low-cost housing bill "to air-condition slum dwellings":

Air conditioning plus television is almost a sure-fire bet for this purpose. I realize that such an approach to the solution of a social problem of immense proportion sounds awfully pat or even cynical or frivolous.

Yet, I think it deserves serious thought, since in some sense it goes directly to one of the roots of the trouble – personal discomfort.[29]

Four days later he wrote to John McCone (1902–91), former head of the Atomic Energy Commission, director of the CIA during the Kennedy administration, and chair of the riot investigation committee in California. Weinberg again pitched "a sort of 'engineering' approach to a partial solution to the problem – that is, simply, provide air conditioners in slum dwelling areas," later estimating the cost as $40 million for some 150,000 dwellings.[30]

Whether the letters' recipients judged "personal discomfort" as important a root as rioters' anger about endemic racism, poverty, and lack of opportunity and self-direction was moot; no written responses are extant. Weinberg nevertheless more cautiously pursued the idea with Emmanuel Mesthene that autumn. "I am increasingly impressed with what I call the 'Cheap Technological Fix' as a means of circumventing social problems." He continued: "a somewhat fanciful" fix "had occurred to me reading about the Los Angeles race riots":

> Someone showed me Huntington's correlation between the incidence of hot weather and race riots in India, and this immediately suggested that providing air conditioners ... might considerably reduce the probability of race rioting. Admittedly this is a superficial and possibly heartless approach to the problem; yet it has the advantage that it just might work.[31]

In the letter, Weinberg introduced two further examples of Cheap Technological Fixes to Mesthene. The first concerned intrauterine devices (IUDs) for birth control. He suggested that "before the ring was invented, birth control was a desperately complicated social problem – its solution required convincing many, many individuals to change their habits and their outlook" but that the IUD "greatly simplifies the matter" by "drastically reducing the daily motivation" needed for birth control.[32]

Similarly, he identified nuclear weapons themselves as a Cheap Technological Fix. "Before the H-bomb," Weinberg noted, "the problem of war was largely viewed as being insoluble unless we changed 'human nature,'" but "by exploiting the crassest notion of self-preservation, the H-bomb offers a quite different 'solution' to the problem of war than the whole Judaeo-Christian tradition teaches is possible." He mused enthusiastically that such fixes could motivate an evolution from social problem-solving towards technological methods, and he recommended it as a line of enquiry for Mesthene's new Program on Technology and Society:

> So to speak, to push the theme to its illogical end, one can ask "will technology make social science obsolete?" – i.e. "can a cheap technological fix be developed for every social problem that short-cuts and makes irrelevant the issues of human conflict that underlie the problem as traditionally viewed?"[33]

Weinberg's hopes thus went far beyond the stopgap of literally cooling down ghetto tensions to provide time for thoughtful social solutions: they suggested long-term technological interventions that permanently *bypassed* sociological approaches, public education, diplomacy, and, indeed, religious and moral teachings. And Weinberg's example of the IUD for birth control (his wife was active in the Planned Parenthood movement) hinted at a further quality of Cheap Technological Fixes: they might work best when they shifted power towards technologist problem-solvers and away from more culturally bound recipients. Unlike the daily personal regimen of the recently introduced birth control pill, IUDs required infrequent attention and always in association with a medical authority. Such examples evade labelling as either naïve or glib. Instead, I would suggest that they illustrate strong faith in rational problem-solvers and attention to aspects amenable to rapid improvement. He was pleased to find that the idea "rang a bell" with Mesthene and, while claiming not to have "the time or knowledge to push the matter," soon invited him to Oak Ridge and the Cosmos Club.[34]

The second timely trigger for Weinberg's private reflection about fixes was the contemporary campaign for automobile safety championed by attorney Ralph Nader (1934–).[35] As he intimated in a "fan letter," Weinberg had been struck by Nader's avowal on the TV program *Meet the Press* of being "interested in remedies, not causes." Weinberg identified his own interest in "the broader question (of which auto safety is one instance)" and enclosed the text of "a couple of talks on the subject":

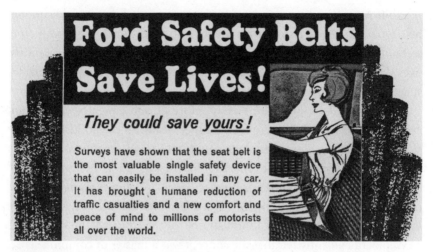

Figure 4.4 Seatbelts as a quick fix for driving dangers.
Ford Australia advertisement, 1964.

> I mention your approach to auto safety as an instance of technology
> circumventing the social and psychological aspects of problems that
> are normally thought of as primarily social rather than technological.
> I originally gave the talk with a bit of tongue in cheek, but as I see
> the logic of your approach to auto safety, I begin to think that tech-
> nology can replace social engineering to a much greater extent than
> I had suspected.[36]

Weinberg had discussed the campaign with Nader's sister Claire (1928–), an
ORNL sociologist with whom he interacted regularly as part of the lab's civil
defence research project.[37] His "talks" were two high-visibility speeches, and
they suggest Weinberg's evolving confidence in the technological fix. The
first had been conceived in February 1966 for a lecture in memory of Alfred
Korzybski. Introduced by Eugene Wigner, the April speech was entitled "Will
Technology Replace Social Engineering?" The second, six weeks later, was
his last-minute choice for an alumni award speech at the University of
Chicago, and it was more diffidently entitled "*Can* Technology Replace Social Engineering?"[38]

While Alvin Weinberg circulated his University of Chicago speech among his acquaintances, he also actively sought its publication. In essay form, the technological fix speech was reprinted in no fewer than seven periodicals over the following year, ranging from journals of behavioural science to space engineering.[39]

More targeted variants of the essay informed subsequent speeches and articles. In late 1966, Weinberg focused specifically on the political dimensions by exploring how technology could stabilize international relations. He also sought more popular venues for his message, ranging from his local Oak Ridge newspaper to the *Los Angeles Times*. Weinberg's audiences grew correspondingly: that autumn, he fielded feedback from a local minister, a law professor, and a medical administrator for the United Mine Workers.[40]

Weinberg evidently felt that the concept was a matter of common sense yet difficult to pin down. To Harvey Brooks he confided his aim to identify a general method of identifying technological fixes: "It would be a neat trick if the social problems could be converted into technological problems as, for example, Ralph Nader is converting the problem of auto safety."[41] Others had sought the magician's trick of transforming the social into the technical realm. Brooks was probably responsible for pointing Alvin Weinberg to the publications of Richard Meier, explored in chapter 3.

Weinberg soon discovered the pertinent echo of his own views in Meier's writing. Discussing social problems, Meier had observed: "Most people would instinctively exclude the scientist and the technologist in the search for solutions. Yet, in many instances, a social problem can be restated so that it is also a scientific or an engineering problem that is not only researchable, but soluble!"[42] Writing that "this is very much in agreement with my current views," Weinberg invited him to lecture at Oak Ridge, an offer that Meier apparently never took up.[43]

Ralph Nader responded more positively to Weinberg's essay: "the points and address which you enclosed were rapidly absorbed." He invited Weinberg to a conference on unreported problems and "a panel on nontechnical obstacles to innovation" in which "the key strategy of the technological fix" would be discussed. Like Brooks, Nader traced Weinberg's notions to antecedents:

The illustrations of the remedial "fix" have been with us for years everywhere, from the automatic coupler on the railroads to a stairway railing, to a machine guard. Now, when it comes to the 60-year-old, man-machine relationship of driver and car, the idea has to be presented as if it is de novo.[44]

Nader's examples were, in fact, remarkably similar to those cited in speeches and literature by American Technocrats from the 1920s through the 1960s. The speeches of Howard Scott and the *Technocracy Study Course* written by M. King Hubbert had discussed examples of guard rails and doors on trams and trains, machine guards on factory equipment, and even rational road design and lighting to direct human behaviours without the need for fines, rules, or training.[45] There is no evidence that Weinberg had encountered Technocracy literature before he formulated his notions of the technological fix, but he had met and corresponded with Hubbert about energy matters in 1961, and Hubbert later remarked on the similarity of their views.[46]

A Catechism for Technological Faith

Ideas first popularized by the Technocrats are readily recognizable in Alvin Weinberg's discourse. Weinberg had described himself early in his career as "a Progressive" who was concerned with the social responsibility and benefits of modern science.[47] In Weinberg's words, "A technological fix is a means for resolving a societal problem by adroit use of technology and with little or no alteration of social behavior" and "technical inventions that could help resolve *predominantly social* problems."[48] His complacent views mirrored the self-pride of the Technocrats and their faith in a society managed by technological innovations, and in which "social problems could be converted into technological problems."[49]

The similarity of these articles of faith for Scott and Weinberg, first introduced in chapter 2, is suggested by a comparison of their respective rhetoric in public addresses:

(a) Social problems of modern society are caused, and ultimately solved, by technological change
As director of a national laboratory dedicated to the development and application of nuclear energy, Weinberg was unusually frank in his acknowledgment of side effects from modern technologies. An admirer of Rachel Carson's work on the unintended ecological side effects of DDT, he was also

vocal about the unsolved problems of nuclear waste and endemic safety issues with reactor designs and management. Yet Weinberg remained an ardent optimist about the powers of technological innovation to solve societal problems and to do so more effectively and speedily than conventional social and educational approaches or via political or religious ideologies.[50]

(b) Conventional solutions – notably economics, politics, and social programs – are ineffective
During the financial crisis of the early 1930s, Scott had been careful to distinguish the views of Technocracy Inc. from radical politics. This was reiterated periodically in the organization's rhetoric, particularly during the Second World War, when the Canadian government banned the organization because of its perceived anti-war stance, and after the Second World War, when the Red Scare discouraged political and social nonconformity in North America. Instead, Scott argued that both the "Price System" (capitalist economics) and communism were outmoded and that "Marxian political philosophy and Marxian economics were never sufficiently radical or revolutionary to handle the problems brought on by the impact of technology in a large size national society of today."[51] In their place, he proposed technocratic problem-solving to circumvent politics entirely: "Technocracy has proposed the design of almost every component of a large-scale social system … a technological socialization is far more reaching, more drastic and more pervasive than anything that Marx or any socialist ever thought of."[52] Weinberg, in his published 1966 speeches mooting the value of technological fixes to replace "social engineering," made similar claims.[53] The "Marxian view," he noted, "seems archaic in this age of mass production and automation":

> Technology has expanded our productive capacity so greatly that even though our distribution is still inefficient, and unfair by Marxian precepts, there is more than enough to go around. Technology has provided a "fix" – greatly expanded production of goods – which enables our capitalistic society to achieve many of the aims of the Marxist social engineer.

Like Scott, he argued that technological solutions could deliver a progressive society "without going through the social revolution Marx viewed as inevitable."[54]

(c) Rational technological change of environments can produce new social behaviours rapidly

A recurring theme in Howard Scott's public addresses had been the rationale of public safety. A rare departure from his streetcar example was his more general claim concerning how safety can be designed into technologies. In an interwar speech, he derided the upsurge of "propaganda against auto accidents" and "great safety campaigns":

> Yet accidents increase. Why? Not because drivers do not know how to drive. With split highways, one-way traffic, raised or sunk crossings, poles and trees moved back fifteen feet from the road, with 4 degrees horizontal and 4 degrees vertical curves only, and with the highways lighted by sodium lamps, accidents would be reduced by more than 90 percent.[55]

The precise figures provided a reassuring counterpoint to emotive calls for public education. Behavioural change through public advertising was ineffective, he suggested, not merely because of recalcitrant citizens but because the modern world placed unfamiliar demands on non-experts. Some thirty years later, Weinberg recounted a remarkably similar example borrowed from Nader: "a safer car ... is a quicker and probably surer way to reduce traffic deaths than is a campaign to teach people to drive more carefully."[56] Weinberg reinvigorated the potency of Scott's and Nader's railroad, factory, and car safety examples by providing more dramatic illustrations of what he described as "Quick Technological Fixes for profound and almost infinitely complicated social problems, 'fixes' that are within the grasp of modern technology, and which would ... eliminate the original social problem without requiring a change in the individual's social attitudes."[57]

(d) Only technically competent people, by redesigning physical environments, are equipped to solve modern social problems

Both Scott and Weinberg broached this claim indirectly. It had been overt in the writings of Thorstein Veblen, who had argued in 1921 for a "soviet of technicians" as a body of experts to govern society.[58] Scott, however, consistently shunned such a prediction and, instead, argued for special qualities of engineers that transcended politics, social policy, and ethical orientation:

> Engineers do not disagree on facts. They all know which direction a stone will drop. They all know that a straight line is the shortest distance between two points. If there is anything else they want to know as engineers, they find it out; and when they find it out, there isn't the slightest disagreement ... As engineers, they are no more radical than a yardstick and no more conservative than so many degrees Fahrenheit.[59]

Instead of a revolution that would replace politicians with technologists, Scott claimed that engineering methods would inevitably be recognized as the *only* means of exerting order in the modern world. As the "application of the knowledge of science and the methods of technology to social management," Technocracy would provide "a blueprint for the operation of the North American Continent, in the same way that an engineer draws up the blueprint for an engineering project, or for a new design of engine, or for a continental telephone system":

> In doing this, Technocrats are not ... influenced by any ethical ideal, but they are primarily concerned with function. The question in their minds is this: What design of social mechanism will operate at the maximum of efficiency with a minimum of oscillation?[60]

A generation later, Alvin Weinberg was even more cautious. His established role as director of a national lab may have provided insights and constraints unfamiliar to Howard Scott. Weinberg seldom broached a direct criticism of politicians, but he repeatedly turned his sights on social scientists:

> The technologist is appalled by the difficulties faced by the social engineer; to engineer even a small social change by inducing individuals to behave differently is always hard even when the change is rather neu-

tral or even beneficial ... By contrast, technological engineering is simple: the rocket, the reactor, and the desalination plants are devices that are expensive to develop, to be sure, but their feasibility is relatively easy to assess, and their success relatively easy to achieve once one understands the scientific principles that underlie them.

Unlike the Technocrats, who eschewed politicians, Weinberg argued for a revision of the working environment that he knew best, as described below. He urged the American government to "deploy its laboratories, its hardware contractors, and its engineering universities around social problems."[61] The physical scientists and technologists who had so effectively conducted the Manhattan Project and postwar development would be reassigned to tackle societal issues via technological methods. Clever engineers would serve as tools of government rather than as replacements for it. Technical specialists would supplement, if not entirely supersede, legislators and educators.

Such comparisons show that Scott and Weinberg shared similar rhetorical ploys, and a significant subset of intellectual convictions. The content, form, and tone of their messages were distinctive. While both typically sprinkled their engineering discourses with technical detail, their presentations to broader audiences instead gave descriptions of technological fixes in the form of generic parables trimmed of context. The compact nature of the brief anecdotes and easily absorbed imagery promoted their effectiveness and retention, and presented an appeal to common sense. The effectiveness of technological solutions to social problems, they claimed, was a self-evident truism.[62]

Turning Technological Fixes into Big Science

As with his promotion of previous issues, Alvin Weinberg interwove his public addresses and articles with writing for advisory committees. His first notions of Big Science coincided with membership on a National Academy of Science committee examining "Basic Research and National Goals." The NAS advised on the level of government support needed to maintain an international lead and the current balance of support to various fields – topics into which Weinberg's Big Science musings fit perfectly.[63] Through 1966, Weinberg dedicated his spare hours to preparing an essay collection, *Reflec-*

tions on Big Science, in which the relationship between social and technical problem-solving mooted the concept of technological fixes. As one reviewer noted, "these reflections represent years of essaying answers to fundamental questions about the changes in science – changes not only in content alone but in scale and scope, in method and purpose."[64]

A subsequent NAS study chaired by Harvey Brooks on technological progress incorporated Weinberg's further thoughts on technological fixes, allied with a shifted focus towards "interdisciplinary, mission-oriented laboratories and the new concern with social problems." This "extension and elaboration of views" from his University of Chicago speech developed the notion significantly: he now conceived reconfiguring labs like ORNL into "socio-technical institutes" devoted to solving national problems via a combination of social research and technological fixes.[65]

Its opening salvo was bold: "a technological invention is easier to make and put into use than is a social invention." If technological components in "social" problems could be defined – "if, for example, they find their expression in the invention of a single device – then … the underlying social problems become more tractable." Weinberg rehearsed the now familiar examples of car and traffic safety and the H-bomb as peacekeeper, but he also identified fossil fuels and nuclear desalination as technologies having dramatically negative and positive social implications, respectively.[66]

As to "problems that are much more obviously social and that seem to have very few technological components such as crime, or race relations, or urban development," he argued that technical solutions were within grasp. Citing Meier, Weinberg observed that many problems traditionally viewed as mainly social possess, in fact, technological dimensions. For instance, electronic burglar alarms "would considerably increase the risk a prowler would have to accept in accosting his intended victim" even if they did "nothing to eliminate the causes of crime: poor environment, poverty, broken homes, and the like." Such "technological palliatives, or even 'fixes'" could be sought more systematically, he argued. "There is a severe mismatch between the Government's magnificent scientific resources for attacking *technological* problems and the seeming *social* character of the problems that the Government is trying to solve." Weinberg consequently proposed that "the country's technologically oriented … laboratories and hardware contractors" be "modified and mobilized to find partial solutions to deeply important social problems."[67]

Weinberg's contribution to the NAS report was reproduced in the ORNL *Review* and distributed to his peers, including an academic in the social sciences at Harvey Mudd College, with the "hope that your freshmen will tear it to pieces as only 19-year olds can." His recommendations were championed by Tennessean senator Howard Baker Jr (1925–2014), who called for a Senate select committee on technology and the human environment and "a dialogue between social scientists, technologists and other experts about national problems."[68]

Alvin Weinberg's new focus confronted national policy directly and proffered technological fixes as a political tool. Two years after first floating the idea, he lobbied the Johnson administration for air-conditioned low-income housing, garnering coverage in the *Washington Post*. And writing more ambitiously to the Department of Defense in early 1967, he suggested that constructing a wall between North and South Vietnam would reduce incursions from the north, a suggestion echoing the contemporary proposals tabled in the President's Science Advisory Group by the Jasons and the Cambridge Discussion Group. Weinberg noted that ORNL "had done a little thinking about it" and "would be ready to mobilize around the problem." Soundings among his peers may have moderated his confidence, though: the following month he backed away from an article on the idea, describing the notion as "very amateurish," and deleted the example from the draft of a forthcoming college commencement address.[69]

In the place of such contentious technological responses, Weinberg offered an example that more potently illustrated the role of renewed national labs in solving socio-economic problems: the development of nuclear agro-industrial complexes. The concept owed much to Richard Meier's work a decade earlier, which had illustrated how complex technological systems could improve underdeveloped regions. It also evoked the experience of the State of Tennessee, home to Weinberg and his ORNL: the Depression-era Tennessee Valley Authority (TVA) project had ambitiously combined hydroelectric dams, waterway diversions, chemicals factories, and distribution networks to provide irrigation, flood control, fertilizer production, and electricity – indeed, the electricity that later had powered the uranium separation plants at Oak Ridge itself.

Alvin Weinberg envisaged vast nuclear power stations as the hubs of such networks. They would generate copious electrical power to desalinate seawater, energize irrigation systems, manufacture heavy chemicals, and provide the motive force for an industrial society. The idea shifted the technological fix from the notion of a short-term repair to a tool of international development, as Weinberg argued at a Swedish conference.[70] It also updated and generalized an ORNL research project, spawned by the Eisenhower administration's "Atoms for Peace" initiative of the late 1950s, to investigate nuclear desalination plants for supplying water for arid regions in the United States. Weinberg's vision consequently pulled together his experience as lab director, essayist, and government advisor.[71] As Weinberg later recalled, "I regarded nuclear energy as a magical panacea … [with] seemingly unlimited possibilities … for solving social problems, poverty, ethnic rivalries exacerbated by quarrels over water, even war itself."[72]

The initiative, developed principally by Lewis Strauss and Weinberg in collaboration with Israeli and Egyptian engineers, was not pursued by the Johnson administration. Weinberg subsequently declined an offer by Strauss to join a Richard Nixon campaign group as he "assiduously tried to separate" his "personal political beliefs from public statements." Instead, he sent a briefing paper to each of the major presidential candidates describing agro-industrial complexes as the "Apollo of the 70s." He argued that federal funding was crucial for such technology projects that were "too expensive, too long-range and too important for the long-term future of the country to be supported by the free market." From technologist-administrator and consultant, Weinberg had moved to a position of direct political lobbying.[73]

Contemporary Critiques and Deft Defences

Alvin Weinberg's networking, speeches, and widely reproduced essays garnered varied attention for technological fixes. Significantly, his closest confidantes – Wigner, Brooks, and de Solla Price – never adopted the term, and Mesthene limited his public acknowledgment to including Weinberg's University of Chicago essay in a 1967 anthology. By contrast, Weinberg's direct superior, AEC director Glenn Seaborg (1912–1999), provided an early endorsement, and University of Chicago sociologist and editor of *Minerva* Edward Shils (1910–1995) built an enduring relationship, drawing on Weinberg as consultant, referee, reviewer, and article-provider.[74]

Weinberg's rhetoric provoked and yet attracted audiences, who appear to have found the notion of technological fixes variously compelling, naïvely confident, or threatening. The earliest disciplinary criticism came from scholars who disparaged Weinberg's phrase "social engineering" – a term that he had employed in correspondence from at least early 1966. Among the first such criticism was that from a representative of the American Geophysical Union, who identified the phrase as an aspersion on engineers. Weinberg replied to suggest a different target:

> The whole burden of my article was to point out that the technologist – i.e. the real engineer – has much to offer in the solution of social problems that are usually considered to be the province of those who try to *manipulate* social behavior; it is the latter whom I call "social engineers."[75]

This reduction of the social sciences and traditional humanistic approaches for problem-solving – including education, politics and ethics – to mere "engineering" and "manipulation" provoked his counterparts in those fields. Even worse, Weinberg had repeatedly dismissed such techniques as inferior to technological innovation. Sociologist of science Bernard Barber (1918–1996) dubbed Weinberg (and his ally, Harvey Brooks) a "gifted amateur" and "scientist-sage" but argued: "science policy studies need to keep close contact with the fundamental social science disciplines, with their best theories and their best research methods and findings."[76]

Richard Meier criticized Weinberg's proposal to scale up technological fixes in agro-nuclear complexes, an idea that he had explored himself. Such "nuplexes," Meier claimed, were a variant of earlier optimistic sociotechnical systems; the history of creating such large conglomerations of interdependent ancillary industries had shown them, he argued, to be "dismal failures" because they had not addressed "people problems" relating to "settlement procedures, laws, customs, education, marketing, management, and so on." Meier underlined that, in order to be successful, technological fixes required carefully planned social interventions.[77]

The most direct appraisal appeared nearly a decade after Weinberg's first speeches on the subject. Sociologists Eugene Burns and Kenneth Studer characterized Weinberg's notions as simplistic. "It is not difficult to see the immediate sanity of such technological fixes as safer automobiles and polio vaccine, for these fixes complement many traditional social values. It is quite another thing, however, to believe that poverty is anywhere near as tractable as Weinberg suggests":

> His insensitivity to the *social* structure of social problems (e.g. with regard to poverty, relative deprivation, rising expectations, etc.) only too readily reveals the technico-scientific bias of his solutions ... Weinberg is thereby constrained to focus primarily on second-order social problems, namely problems that are precipitated by prior technological fixes.[78]

In a subsequent exchange, the authors summarized Weinberg as proposing "scientistic, reductionistic, solutions" – an unsupportable faith in the methods of physical science and a myopic approach to problem-solving. Weinberg disingenuously denied "any grand social philosophy" or "mutual consistency" in his writings. "The authors read too much into my views when they claim to see this as evidence of my naïve belief in the possible redemption of society by science." In a more convoluted defence, he observed that "social fixes, no less than technological fixes, have deleterious and unforeseen side effects; and social fixes, precisely because they get to the heart of the matter rather than remedying effects, have a history of going more awry than do technological fixes."[79] Critiques of technology, like expressions of technological faith, grew in a particular historical context. As Weinberg promoted his views during the late 1960s and early 1970s, his critics identified reliance on technological solutions as evidence of inadequate engineering practice, failures of government policy, or expressions of corporate self-interest.

As early as 1970, Weinberg attacked technological pessimism by appealing in his commencement addresses to college engineering graduates. For these audiences, Weinberg's defence was nuanced. He acknowledged that "technological fixes are viewed with great suspicion ... by many social activists" and that "we technologists are aware of the shortcomings"; he noted that the Green Revolution in India had "created unemployment among agricultural workers as well as undermined the social structure of the village." But, more directly addressing politics, he claimed that "social fixes" had a history of errors at least as great: "our US Constitution, with its intricate and almost mechanical checks and balances, represents a social fix on a grand scale. Or

Karl Marx, with his curious mixture of Hegel and science, created an all-encompassing system which was supposed to lead to Utopia on Earth." And, hinting at the ongoing impeachment proceedings against Richard Nixon, Weinberg underlined: "our US Constitution is imperfect; the incredible stresses it is now undergoing are evidence enough of this." In its place, he argued, technology could serve as a calming counter-force to unstable politics: "Our political system is geared to a four-year cycle. This our founding fathers recognized as necessary to prevent against tyranny. But in guarding against tyranny we have created a mismatch: the time scale of politics is far shorter than the time span of technology." In effect, technology served as an automatic governor – a conservative counter-force to damp out damaging social oscillations.

The "new style of social thinking," he suggested, "forces us to look at our socio-technological dilemmas from a longer perspective of time than is our custom." Sound technological optimism, Weinberg concluded, tipped the balance from social to technological fixes: doomsayers quoting *Limits to Growth* would be countered by young technologists "who foresee solutions to all its problems." This placed a unique responsibility on the technological elite to "use this sophistication to break prevailing chains of short-sightedness – even though at times it may place you in temporary disfavor. Your technological fixes will thereby become more humane; and, more importantly, each of your lives will thereby be made more richly human." Thus engineering paternalism, not democracy, suited the troubled modern world.[80]

The political dimensions of technological fixes elicited perhaps the most enduring interest of critics and illustrate the evolution of Weinberg's views concerning technological power and expertise. As early as 1966, a reviewer for the Socialist Labor Party described Weinberg's thesis as "speciously attractive" and agreed that "social engineers [are] woefully ineffective," but it identified Marxism as the only genuine social science. Identifying technological change as "the force to which society's institutions must ultimately conform," it castigated Weinberg for having given up his conviction that society's agenda should focus on the major task of social reorganization.[81]

Others argued that the technocratic faith implicit in Weinberg's work required an unlikely and worrying transfer of power to technologists, a criticism levelled at Richard Meier's writings a decade earlier and, indeed, at the

Technocrats of the early 1930s.[82] In his lobbying for socially oriented Big Science, Weinberg had recognized that think tanks of experts were potentially dangerous because the technical complexities of the issues prevented public debate. "Many of our strategic doctrines ... can be traced to RAND [and] it is somewhat disconcerting that they are formulated by experts who, at least from the outside, appear to sit apart and to operate on their own." His proposed solution was to establish two competing institutes, which, like the Oak Ridge and Argonne, and Los Alamos and Livermore, national laboratories, "[would] keep each other honest." In this respect, Weinberg's views deviated from those of the earlier Technocrats. Where they had proposed replacing political leaders with governance by technologists, Weinberg envisaged engineers and scientists in government-funded national labs devising solutions to problems identified by others. Both views, however, circumvented direct democratic involvement in favour of rational elites.[83]

Ethicists, too, weighed in. Max Oelschlaeger (1943–) argued that the notion of the technological fix had become a popular myth adopted by large companies even more than the general population.[84] The ethical ramifications of technological reliance were more centrally addressed by philosopher Arne Naess from the early 1970s, as discussed in chapter 8. He argued that the seeming "reasonableness" of these solutions was largely determined by the narrow framing of the problem and failed to explore the cultural presuppositions about the nature and potency of technologies. Naess suggested a new framing in which technologically oriented, short-term *shallow* environmental solutions would be replaced by *deep ecology*, which would seek to address systemic faults holistically via a combination of social, cultural, and technical solutions.[85]

During the 1980s and especially via Edward Shils and *Minerva*, Alvin Weinberg continued to contribute to science and technology studies and reflected on their political dimensions. He judged Langdon Winner's piece, "Building the Better Mousetrap," to be "the most readable and cleverest of essays" and admitted learning more about the political ramifications of technology from his "graceful essays" in *The Whale and the Reactor*, while seemingly missing the most striking technological fix offered in that book: Winner's description of New York City parkway overpasses built low so that buses carrying urban poor were unable to reach prime suburban recreation areas. Equally surprisingly, there are no extant documents revealing his views about Winner's best-known piece, "Do Artifacts Have Politics?," which explores the link between technological innovation and political effects, a connection sought by Weinberg for two decades. Instead, he proffered that Winner was too pessimistic, suggesting disingenuously that, "as a technologist unversed in politics, I would claim that technological fixes are easier than social fixes." Indeed, in his 1994 autobiography he characterized his career as that of a "technological optimist" and "technological fixer."[86]

Trajectory of a Concept and Career

There are ironies in Alvin Weinberg's promotion of the technological fix. Career experiences repeatedly challenged his life-long faith in technological innovation as a societal resource. Weinberg was an admirer of Rachel Carson and had promoted paying attention to the environmental side effects of technologies. He was an early advocate of nuclear power as a low-carbon fix to avoid climate change, and he regretted having done little at ORNL to focus attention on the persistent problem of nuclear waste disposal.[87] By the end of the 1960s, however, this progressive stance – distinctly out of step with many of his establishment peers – became blurred. His public addresses were increasingly targeted at "primarily the young, anti-technology revolutionaries and their more passive, but worried followers."[88] Yet his outspokenness was not on-message within the nuclear industry either. He was dismissed as director of ORNL in 1973 during the first Nixon administration because of his publicly expressed criticisms of reactor safety.

Fashioning a role as director of the new Institute for Energy Analysis at Oak Ridge, Weinberg's proposed alternatives eroded his complacent championing of technological fixes (figure 4.5).[89] He envisaged a combination of technical and social components to the solution: a new generation of "inherently safe" reactor designs but strategically sited in large clusters far from populations in high-security "nuclear parks" tended by a "nuclear priesthood" of specialists. Ironically, technological improvisation thus ceded place to ponderous societal rearrangements.[90] In a similar vein, he later suggested that the Hiroshima atomic bomb had been necessary and effective in the short term but that only its public "elevation … to the status of a profoundly

Figure 4.5 Weinberg as self-proclaimed "technological fixer" and founding director of the Institute for Energy Analysis, Oak Ridge Associated Laboratories (ORAU), c. 1982.

mystical event" could avoid future nuclear wars. Weinberg's neologism "trans-scientific" labelled such problems that transcended scientific analysis and that required, as corollaries, social and moral considerations.[91]

Over the same period, Weinberg reluctantly came to recognize that his life-long optimism for nuclear power was not shared by the American public. In a letter to Edward Shils, he concluded, "the sorry history of nuclear power seems to support a growing view that risky technology is incompatible with liberal democracy," and, reflecting on the relative success of nuclear programs in other countries, he observed, "nuclear energy seems to do best where the underlying political structure is elitist." Alongside Langdon Winner, he acknowledged that certain technological solutions could not only produce powerful political consequences but might also require particular political environments.[92]

And yet Alvin Weinberg's curiously imprecise conception of technological fixes was successful, remaining in public discourse from its origin in the mid-1960s throughout the following four decades of his life.[93] This chapter illustrates how his notion of the technological fix reflected the views of his peers and career context, and how it can be traced to experiences shared by many American technologists and analysts of his times. Why, then, did his particular ideas create such an enduring impact?

First, Alvin Weinberg's voice was favoured as he was a national laboratory director with privileged access to policy-makers and industrialists. His active networking engaged peers from a broad range of disciplines as critical friends. His career translated him from an interwar "Progressive" to an apolitical champion of progressive technologies. Instead of opposing and seeking to replace politicians as had the Technocrats, he conceived nationally funded techno-fixers and their technological fixes as tools of good government.[94]

Second, Weinberg was a skilled and tireless communicator. He crafted his writings to be readily absorbed and was unusually active in promoting the republication of his essays for varied audiences. Richard Meier suggested that his coinage was seductive, too: the word "fix" connoted "a dramatic improvement, and therefore, it was widely discussed." Other scholars, while critical of Weinberg's views, praised his role as "one of the more outstanding and articulate of the science policy thinkers," whose clearly formulated views allowed "a more incisive criticism."[95]

On the other hand, Alvin Weinberg's shifting public stances effectively reflected his evolving confidences and his varied audiences. He was at his most bullish and optimistic in commencement addresses to college engineering graduates, but he was tentative and conciliatory in communications directed at scholars of social sciences and humanities. Weinberg's musings framed questions to encourage discussion.

Third, and most important for his long-lived influence, Weinberg's rhetorical style eluded detailed unpacking. A "Teflon man" before the term was coined for Ronald Reagan, he achieved fame with little notoriety by deftly shedding criticism. Like those of the Technocrats, his examples were brief and appealed to common sense rather than to analysis. Definitions of technological fixes were mutable, alternately conceived as temporary solutions that bought time for social methods or as large-scale systems that

shaped citizen options and replaced social science altogether. He also skil-
fully pre-empted criticism by raising some of his critics' objections himself.
Playing devil's advocate and portraying himself as neutral and diffident
rather than polemical, his public ambivalence preserved his critical stance.[96]

In later years, Weinberg identified himself more overtly and confidently
with technological fixes. What he had painted as disingenuous "what-if"
scenarios in the late 1960s were openly represented as a matter of personal
conviction and even defiance a quarter-century later. In his final decade,
Weinberg reflected that "Paul Ehrlich's derisive description" of him as "king
of the technological optimists" was justified but that he nevertheless felt vin-
dicated: "I write as the Cold War has ended. Deep in our hearts we realize
that all this was made possible by a technological fix – the hydrogen bomb."[97]

While criticized by some audiences, Weinberg's seductive notions were
mainstreamed to shape the confidences of those seeking solutions to urgent
and enduring societal problems. Notions that had, mid-century, been the
province of a technological elite became a widely shared belief of broader
publics. As an articulate voice for technological solutions to wider societal
issues, he consolidated and lucidly framed ideas circulating among his peers.
By offering examples and discussing their wider implications – and just as
importantly, by labelling the concept concisely – Weinberg's rhetoric shaped,
and continues to influence, discourses about technological solutions to societal problems and about the wider roles of technology in society, as is explored in subsequent chapters.

Revisiting Technocratic Fixes after Howard Scott

Alvin Weinberg's appropriation of the Technocrats' metaphor, recounting
logical anecdotes of engineering wisdom as new parables of the modern age,
brought the notion of technological fixes to receptive audiences at gradua-
tion ceremonies, government policy consultations, and peer conferences.
This new perspective nevertheless echoed technocratic themes. For less elite
audiences, the populist rhetoric of Technocracy Inc. continued to shadow
Weinberg's message throughout the 1960s. Howard Scott's version – that
technological fixing could not be allied with conventional political processes
and, instead, required direct management by untainted technical experts –

was more amenable for disenfranchised publics. His supporters complained that conventional governments, politicians, and economic interests channelled outmoded methods and favoured inappropriate power-holders.

It is ironic, then, that Scott's technocracy, which had touted the improvement of social environments by benevolent engineering interventions, was increasingly criticized as a dangerous trait of modern governments. The term "technocrat" was used increasingly by critics as a pejorative label for a political elite manipulating the mechanics of government and the economy to wield power. This popular understanding of Technocracy – as a reductive and inhuman force within conventional government – was actively but unsuccessfully opposed by Howard Scott's organization after his death in 1970.[98]

The subsequent activities of Technocracy Inc. contributed inadvertently to this alienation from its audiences. Following Scott's death, the organization retained and only cautiously expanded upon Scott's views, somewhat akin to a nascent religious tradition. Reminiscent of the apostolic period of Christianity, the canon of speeches, articles, and pamphlets was collected and reused, to be quoted in relation to contemporary technological and societal issues. The *Technocracy Study Course* was repackaged in abbreviated form to act almost as a catechism for the organization's credo. The old technological examples were recounted but not extended. Instead, the texts were subtly recast in ways that suggest the influence of Alvin Weinberg's independent advocacy. The organization accompanied the term "technocracy" with straplines that highlighted and streamlined its link to societal change. Technocracy was now defined pointedly as "the scientific method applied to sociological problems" and promoted as "the Technological Social Design, permitting science to formulate a scientific socio-economic structure."[99]

And while a seminal successor to Howard Scott failed to materialize, the organization fielded a handful of lesser evangelists for Technocracy and prophecies of imminent societal catastrophe. With regional chapters and membership income falling, Technocracy Inc. increasingly directed its energies towards a kind of missionary work: converting influential public figures thought to have a public identity relating to socially responsible science, futurism, or an anti-establishment orientation. Targets included science writer Isaac Asimov (then representing Zero Population Growth), ecologist-ethicist

Figure 4.6 Rational design to replace political governance. Technocracy Inc. public outreach presentation slides, 1977.

Garrett Hardin (author of controversial essays on managing environmental resources), *Omni* magazine editor Ben Bova, and dozens of others.[100]

Replies to the letter-writing campaigns were rare and unpromising. Aimed at critical public figures rather than broad audiences, the initiative had eschewed the compelling parables and icons for less concrete arguments that had little persuasive power. Garrett Hardin responded, "the problems of human relations and the allocation of scarcities cannot be altered or escaped by worship of technology"; the editor of the *Bay Area Skeptic* retorted, "do not assume that I am interested in exchanging letters on your Utopian Technocracy." Few others replied at all.[101]

A senior Technocrat mourned the cultural decline of their brand of technological faith:

I do feel that Technocracy Inc. has lost its credibility and its appeal especially to people of technical and scientific sophistication, and indeed, to the public at large ... In general it sounds more like a political opposition party, constantly nagging away at the status quo. How tiresome. And how futile. Where is the scientific thrust that focused on the unidirectional and irreversible progression of science and technology that gave rise to the concepts of Scott's Technocracy in the first place?[102]

But while the radical Technocrats failed to attract new followers, Weinberg's vision of technologies for social benefit was attuned to mainstream power-holders and wider publics.

5 Technological Promises in Popular Culture

Viewing Human Ideals through the Lens of Invention

Earlier chapters closely examine how confidence about technological solutions grew within fringe groups and professional communities. This one shifts our gaze. It resets the clock, stepping back in time to trace how technological faith spread from such influencers to become a tenet of popular culture. It also adopts a wider perspective to capture these more diffuse currents in society but with an eye that retains a sharp focus on faith in fixes. Prior authors of cultural history have written compellingly about the popular expression of technological wonder and optimism about progress. I draw upon some of the same sources but pay particular attention to how, throughout the century, popular media expressed growing confidence in deliberately configured technologies and the social benefits that they promised. Less tightly bound to engineering culture, these visions of how science and technology would contribute to the better society were diverse and sometimes conflicting. Fiction portrayed distant forecasts of utopian societies empowered by technology; periodicals anticipated scientific progress having social value; news sources reported imminent technological benefits to daily life.

Forecasting the Rational Future in Fiction

Science fiction became the genre for depicting the benefits of the imagined technological future. But before the category had a name, it had an audience.

Without question, the most influential early paean to technology as the basis of a benevolent society was the utopian novel *Looking Backward*.[1] Published in 1888 by writer and journalist Edward Bellamy (1850–1998), the book imagines the city of Boston a century ahead. It is a synthesis of contemporary political, scientific, and engineering thinking. Its author had spent a year in Europe and, specifically, Germany, then at the forefront of scientific and technological advance. Bellamy's United States was also burgeoning in entrepreneurial spirit but was burdened by the corporate competition and government collusion of Gilded Age America. His writing described an alternative future and offered a template for later self-styled Progressives such as Howard Scott and Alvin Weinberg.

Bellamy's fiction made the case for a rational society intentionally configured by technological solutions. He imagined a world founded on cooperation for the common good and social improvement, coupled with the efficient production of beneficial commodities. Written just a decade after the invention of the telephone, for example, the novel describes how concerts and religious services would be carried to individual homes, extending a sense of community to the infirm and aged. Bellamy also applies scientific notions to explain and resolve social issues. He identifies crime, for instance, as primarily due to the structure and inequities of capitalist society. Free availability of the physical necessities and luxuries of life, he argues, will largely eradicate it. Remaining psychological causes of crime would be addressed by innovative medical treatments. Such notions were to be popular in the rhetoric of the Technocrats in subsequent decades and, later still, in examples of technological fixes envisaged by Meier and Weinberg.

A prototype for science fiction before the genre had a label, Bellamy's novel became a bestseller in North America, Britain, and beyond. It carried forward a version of what Marx and Engels had recently described as "scientific socialism," which predicted societal transformations set in course by new technologies and the triumph of reason.[2] The book's matter-of-fact descriptions portrayed not only pragmatic solutions to present-day social problems but also an imagined route to this benign society of the near future, evolving in lock-step with technological advances. So compelling was the narrative that the book was reprinted by socialist organizations throughout the early twentieth century (figure 5.1). Bellamy focused on his own country, however – giving it an attention much like that of the Technocrats some thirty years later. Sensing that American readers would be alienated by the word "socialism," he employed the term "Nationalism" to describe his non-competitive, technologized vision of the better world. Within a couple of years of the book's publication, hundreds of "Nationalist Clubs" had been formed in American cities. Bellamy himself published the *New Nation* weekly newspaper from 1891. Through it, he sought to guide links between the Nationalist Clubs and the policies of a new national party, the People's Party. The periodical and political movement faltered over the next decade, but Bellamy's final novel, *Equality*, explored his imagined sociotechnical society in greater depth. In it, he again imagined new technologies as inherently liberating: clothing made of paper, for instance, would be inexpensive and recyclable. New communication and transport technologies would promote egalitarian social interaction.[3] The novels, in short, epitomized the frame of technological fixes before the idea had a name.

Where *Looking Backward* can be recognized as foreshadowing the theme of directed social evolution evinced by rational thinkers (later championed by the Technocrats), *Equality* adopts a more populist orientation. The future society of the second novel appears less hierarchically organized than that of the first. Despite its more overt lecturing tone, a common trait of such utopian fiction, it was widely read and discussed. Bellamy's books fostered numerous unofficial sequels, critiques, parodies, and analyses over the following decades. None proved as captivating as his original vision, but the collective outcome was active re-examination of American political, economic, and technological organization, and exploration of the best route to a rationally ordered society.[4]

The novels and stories of English writer H.G. Wells (1866–1946) evoked some of the same ideas but with important nuances. Wells made his name as a writer of speculative fiction, beginning with *The Time Machine* in 1895. The story describes excursions into the deep future and depicts the ultimate fate of the human species in the process. The plot relies heavily on technological details intelligently described and is informed by contemporary scientific perspectives on topics such as evolutionary biology. Like most of Wells's subsequent novels, however, it hints at social and sociological topics such as class and power in relation to new technologies.[5] Through the first half of the twentieth century, Wells shaped popular thought via his visions of future technological societies and their social outcomes. For example, a novel later admired for its seeming predictions of the development of nu-

Figure 5.1 Foretelling the rational future. Book jackets for *Looking Backward*, for audiences of the 1890s, 1920s, and 1950s, respectively.

clear power and its industrial, social, and political consequences was *The World Set Free* (1914). Gaining a reputation as a writer of history after the First World War, Wells consequently saw the forecasts of his "scientific fantasies" gain increasing authority.[6]

Wells's writing cannot be reduced to prototypical technological fixes or even technological optimism: he imagined technologies producing negative as well as positive social products. For instance, his 1899 text, *The Sleeper Wakes*, has a plot similar to *Looking Backward* but describes a dystopian London some two hundred years in the future following a spoiled revolution. Like Bellamy's book, it describes technologies as suppliers of entertainment and pleasure but available only to the privileged few.

Wells's later writing continued to juxtapose technological and social dimensions. The theme of *A Modern Utopia* (1905), for example, concerns how social problems can be managed in a constantly progressing world. In the book, he illustrates the need for direction by an elite to both manage technological progress and to implement social measures to maintain equity. In *Men Like Gods*, written eighteen years later, he again describes a utopian society shaped by technological tools, such as communicators resembling mobile phones with e-mail. Paralleling the predictions of the Technocrats then emerging in North America his imagined world is directed by the new faith in scientific research, which has replaced both politics and religion, but with overt attention to social outcomes.[7] But Wells's attention to social reforms as parallel tools marked his vision as more refined than that of the Technocrats.

His sociotechnical fantasies also clashed with other contemporary views. Thus, the publication of Bellamy's *Looking Backward* had been famously countered two years later by William Morris's *News from Nowhere* (1890). It followed Morris's review of Bellamy's book, taking aim at what would later be identified as technological fixes for physical labour:

> [A] machine life is the best which Mr. Bellamy can imagine for us on all sides; it is not to be wondered at then that this, his only idea for making labour tolerable is to decrease the amount of it by means of fresh and ever fresh developments of machinery ... I believe that this will always be so, and the multiplication of machinery will just multiply machinery.[8]

The competition of sociotechnical philosophies remained just as vital for Wells a generation later. *The Shape of Things to Come* (1933) was his response to Fritz Lang's film *Metropolis* (1927), a dystopian vision of a stratified society. Wells provided a fictional account of the future, informed by his previous works on history and contemporary political economy.[9] In it, he chronicles an imagined European society between 1933 and 2106. He describes the collapse of the economy and a continent-wide plague, followed by the rise of a benevolent dictatorship. The actions, buttressed with the power of air forces, sound much like the forecasts of Howard Scott's Technocrats: the new leaders – a scientifically trained elite – would implement a technologically managed world society, eradicating religion and economic speculation. Written at the height of the Great Depression, the novel offered an alternative history of the future in which scientific logic would triumph over the failures so evident in the contemporary world.[10]

Adapted by Wells for cinema in 1936, *Things to Come* was a high-budget British film with dramatic special effects that communicated a more technocratic subtext, particularly when divorced from the more overt socialist tone of the novel. Beginning with the coming world war anticipated by its audiences, it depicts subsequent decades of plague, anarchy, and petty dictators. The arrival of the Air Men, engineer-representatives of a shadowy organization called Wings Over the World, heralds a new rational society. Their model is oligarchy: governance by a few wise men. Empowered by their rational organization, this elite oversees the rebirth of industry and social cooperation to create a gleaming future organized on scientific lines. Cities are largely underground and climate-controlled; production is automated; enormous telescreens in public places broadcast the speeches of the beneficent leaders (figure 5.2). This utopia tolerates its discontents, too: a minority of rabble rousers portrayed as artists, artisans, and non-conformists protest the launching of a rocket-ship to the moon as a symbol of pointless progress. Their opposition inevitably fails and, as the rocket is launched, the film ends with a monologue vaunting endless human progress through science delivered by one of the principal characters, Oswald Cabal, a man of science in the technocratic mould:

> For Man, no rest and no ending. He must go on, conquest beyond conquest. First this little planet with its winds and ways, and then all the laws of mind and matter that restrain him. Then the planets about him and at last out across immensity to the stars. And when he has conquered all the deeps of space and all the mysteries of time, still he will be beginning ... If we're no more than animals, we must snatch each little scrap of happiness and live and suffer and pass, mattering no more than all the other animals do or have done. Is it this, or that? All the universe, or nothingness? Which shall it be, Passworthy? Which shall it be?[11]

By contrast, Lang's film depicts technologies as reinforcing social injustice: an industrial near-slave class, living below ground, is ruled by capitalists in airy high-rise homes. It is notable that *Metropolis* reinforces some of Wells's themes, especially the power of technologies to shape social conditions, but suggests that the (negative) outcomes can be resisted only with great effort. The plot is, in fact, an acknowledgment of intrinsic faults of sociotechnical systems and a challenge to technological agency, a perspective that gained wider credence only a half-century later. Similarly, Aldous Huxley observed that his *Brave New World* (1932) was meant to be a revolt against "the horror of the Wellsian Utopia" as depicted in books such as *A Modern*

Figure 5.2. Rational world governance: scene from *Things to Come* (1936).

Utopia and *Men as Gods*.[12] Huxley's novel famously attacks the sterile ratio-nality of a technologized future society. In it, technology serves explicit social ends. Different forms of human are grown in artificial wombs to supply so-ciety with workers of different social niches; drugs supply basic needs and pleasures; education is via sleep-learning.

The proliferating texts in this clash of faiths reveal distinct beliefs about how technology could relate to society. Bellamy's and Morris's novels battled over the essential contribution of technological improvements. Wells's own vision of beneficent elites sensitive to social justice owed much to his dal-liance with Fabian socialism and later interest in Soviet communism. Huxley satirized technological fixes the very year that the American Technocrats achieved their greatest public exposure. And each nuanced their views in relation to the colder rational sympathies of the American Technocrats.

Seeding Enthusiasms: Science Fiction Fandom

The writings of Edward Bellamy and H.G. Wells describe forecasts of technologies directed to achieve deliberate social outcomes. This was a potent theme, especially in the American context, where inventiveness and social consequences seemed closely linked. Technological skills and pastimes had become an increasingly visible activity from the early nineteenth century and publications communicated them to growing audiences. Popular science and engineering periodicals proselytized the values, methods, and (most ardently and consistently) the practical products of the modern world.[13] American science magazines vaunted specifically utilitarian and economic dimensions. *Scientific American* was born in 1845 to capture this public enthusiasm, chronicling new inventions week by week and, later, on a monthly schedule. Over the following seventy years, it was joined by a growing number of popular technical periodicals that conflated scientific discovery, invention, and social benefits.

Popular writing constructed a specifically American identity for this technological enthusiasm. Adolescent fiction in the first two decades of the century equated the mastery of technology with social power. American technology was infused with science; it was active, innovative, and profitable. The *Tom Swift* series of books (1910–41) devised by American writer and publisher Edward Stratemeyer (1862–1930) focus on a young inventor and his adventures with exhilarating electrical and transport technologies. The Stratemeyer Syndicate churned out mysteries that mixed invention, clear thinking, adventure, wondrous capabilities, and industrial secrecy, usually with boys as protagonists. Mirrored by other publishers, several thousand titles provided role models for three generations of American children and young adults.[14]

Some of those same audiences were further inspired by magazines dedicated to hands-on experimentation and innovation. Another seminal American publisher was responsible for a large fraction of these ventures. Hugo Gernsback (1884–1967), an entrepreneur in the early American radio industry, chronicled invention through his periodicals aimed at technical amateurs and emerging science fiction enthusiasts. He followed his first magazine, *Modern Electrics* (1908), with dozens more seeking to capture a

growing public appetite for popular science and invention. In the periodicals, scientific curiosity was blended with technological enthusiasms and
individual expertise to generate new pastimes and potential career skills.
The content and themes of such publications altered markedly after the
First World War to overtly encourage amateurs. In the postwar environment, new publishing initiatives, including a renewed *Scientific American*,
were oriented towards articles displaying more explicit scientific content
and aiming to promote active engagement by enthusiasts.[15] During the
early twentieth century, then, "science" was broadly construed by American
readers of popular literature as what today might be labelled techno-scientific faith: a progressive and culturally transformative activity linked with
personal improvement, economic benefits, and expanding knowledge.

Gernsback's *Everyday Mechanics* (1915–16), for instance, included articles
and colourful cover art that depicted scientific experiments, and *The Experimenter*, subtitled "*Electricity – Radio – Chemistry*," specialized in articles
providing hands-on projects to build and use scientific apparatus. Through
the 1920s, popular titles mutated to reflect science-oriented content more
explicitly. Thus, Gernsback's *Practical Electrics* (launched 1921) became *The
Experimenter* from 1924; *Electrical Experimenter* (launched 1913) became *Science and Invention* from 1920; *Everyday Mechanics* was relaunched as *Everyday Science and Mechanics* (1931). In distinctive ways, such periodicals spread
the new twentieth-century vision of technologies for personal empowerment and social betterment.[16] This popular literature nevertheless mixed
together more than mere faith in technological fixes. If a common philosophy can be discerned at all, it is that new technology automatically produces
social outcomes. This implicit faith in determinism contrasts with the intentionality of the Technocrats' vision of the designed society.

Gernsback's periodicals were seminal in promoting broad technological
enthusiasms. He not only popularized the new breed of magazines for tin-
kerers but also defined a new genre. During the 1920s, such speculative writing gained a name: "scienti-fiction" and, soon afterward, "science fiction."
Interestingly, Gernsback's dogged quest to capture new magazine reader-
ships led him to launch *Technocracy Review* in 1933, at the peak of the organization's popular interest. As a publishing entrepreneur, Gernsback was
diffident about Technocracy's monetary notions based on energy usage. On
the other hand, he identified other technocratic themes – social solutions

through technology and governance by technologists – as the bread and butter of science fiction. Perhaps to cross-fertilize magazine sales, he dedicated
his 1933 editorial in *Wonder Stories* to the "Wonders of Technocracy":

> So far as the aims and aspirations of Technocracy are concerned, they
> are nothing new to science fiction … Indeed, there is very little that
> Technocracy offers that has not been anticipated in stories from H. G.
> Wells down to last month's WONDER STORIES …
>
> Everything the machine age has to offer, many of the possibilities
> arising from the reign of the machine, have been anticipated by authors
> of science fiction for many years. Indeed, it would be interesting to take
> every statement made by spokesmen of Technocracy and check up on
> some of the past science fiction stories.

This was more than mere technological enthusiasm and gadget-wonder.
Science fiction rather than sterile theory, Gernsback suggested, was the way
to explore the exciting powers of technology for society. He criticized not
only the Technocrats' lack of originality but also their weak foundations.
Gernsback identified the organization's leaders as technicians who were
limited to receiving the intellectual products of scientists rather than as scientists themselves. "Scientocracy," he proposed half-heartedly, should replace "technocracy."[17]

Six years later, science fiction writer Ray Bradbury (1920–2012), just beginning his career, adopted a more positive tone. Aged nineteen, he declared
his interest in Technocracy and a few months later dedicated the first issue
of his *Futuria Fantasia* fan periodical to the subject to satisfy "the crying
need for more staunch Technocrats." "Technocracy embodies all the hopes
and dreams of science fiction," he claimed. "We've been dreaming about it
for years – now, in a short time, it may become a reality."[18] To hedge his bets,
Bradbury published a lead article by Bruce Yerke, a sixteen-year-old self-
described Technocrat. It was followed by a satirical story about the likely
effects that a society directed by the Technate might have for hack writers,
allegedly penned by Bradbury himself.[19] Yet the second issue of the mimeographed fanzine had less coverage of the subject, claiming that feedback had
been "funny, if not childish." The two final issues through 1940 promised,
but did not deliver, planned Technocracy articles. Bradbury later claimed

that he moved away from the topic after attending a rally led by Howard
Scott in Los Angeles and noting the similarity between the members' grey
suits, graphic symbols, and jargon and the trappings of fascist governments.
While such writing reached tiny audiences, the example suggests a close association between Technocratic ideas, mid-century science fiction readers,
and would-be writers.[20]

Science fiction writing before and after the Second World War generated
and incorporated such enthusiasts, fanatics, and closet Technocrats. Self-
described "fans" were conduits for spreading these themes. Like Bradbury,
they published fanzines, wrote public letters to editors of commercial magazines, and shared their enthusiasm through clubs and personal contacts.
During this golden age for the genre epitomized by young writers such as
Isaac Asimov (1920–1992), science fiction depicted the technology-driven
future as progressive and inevitable.

Asimov was a member of a New York City science fiction fan group known
as the Futurians, active between 1937 and the end of the Second World War.
To varying degrees, several of its members expressed their political convictions in their writing. Like Bradbury, at least four of them became interested
for a time in Technocracy and, in 1940, took the organization's Study Course.
Within a year, however, they became disenchanted with Howard Scott, a
"large, domineering man" with "[dis]likeable qualities." One recalled going
in as "Stalinists disguised as Technocrats" but winding up as "subdued …
progressive liberals."[21]

The Futurians' aim to politically educate science fiction fans became
known among the fans as "Michelism," described by Sean Cashbaugh as "an
intermingling of Popular Front communism and prevailing views" of science fiction. It was named after member John B. Michel (1917–1969), who
espoused utopian socialist views and the promotion of social progress
through applied science. This was, in essence, a version of Bellamy's script
updated for communication via the new fans of science fiction.[22] Michel argued that science fiction fans were an unusually prescient and active audience and that the genre worked "to produce a certain state of mind which
is destined sooner or later to take a large hand in shaping the destinies of
the world." Cashbaugh suggests that leftist politics motivated Michel and
his contemporaries but hints that technocratic ideals were the basis of a
broader consensus. "Fans saw themselves as an elite cadre of science fiction

literati, typically male, with the unique social authority and power associated with scientific knowledge." More widely shared still was confidence about the dominance of technological drivers for social progress.[23]

As Cashbaugh implies, faith in a utopian technological future was endemic among the first generation of science fiction fans and writers. Although Asimov, a Democrat and liberal, was less overt in his politics, many of his stories carried forward the narrower early twentieth-century convictions about transformative and deterministic technologies.[24] In a series of stories written between 1942 and 1950 and published as a trilogy of books in 1951–53, Asimov developed what could be described as a future history of technological fixing on the grandest of scales.[25] His books, much later expanded into a series of seven, and linked to other series that he had written, painted a panoramic picture of human expansion through the galaxy in the distant future. The principal character is Hari Seldon, a mathematician who advances the field of mathematical sociology (an imagined extension of the real-world research begun by Comte, Quetelet, and Durkheim in the nineteenth century). Seldon develops predictive insights incorporating principles of scientific psychology. The outcome is "psychohistory," a subject allowing the forecasting of broad historical changes millennia in advance. The undercurrent of this imagined subject is the value of scientific rationality applied to societies and determinism in the long-range social consequences of technologies. Highlighting this rationalist worldview, Asimov's fiction typically avoided non-human characters and love interests. In such a fictional context, the logic of necessary actions and their technological effects overcome mere human emotion. For his readers, the consistent and fine-grained scope of the book series suggested a plausible future in which technologies and social effects would unfold in a predictable sequence. Indeed, Asimov aided this coherence by eventually writing prequels and follow-on novels that tie up loose ends of the narrative.

Asimov's Hari Seldon, like Wells's Oswald Cabal, represents the power of scientific analysis and its logical management of society. Both characters argue for the optimistic inevitability of this technoscientific turn and scarcely hint at its fatalistic judgments about free choice and cultural values. In this respect, they mirrored real-world counterparts such as Scott, Meier, and Weinberg.[26]

Showcasing the Inventive Future: World's Fairs and Theme Parks

The rhetoric of World's Fairs converted much larger audiences to faith in the social solutions promised by technologists. The Fairs also provided a direct channel for technology companies to tap into collective dreams. Some two hundred science fiction fans, including Ray Bradbury, Isaac Asimov, and other Futurians, attended the first "World Science Fiction Convention" in July 1939, held in conjunction with the New York World's Fair.[27] The theme of the Fair, "The World of Tomorrow," fitted both the exposition and the fan fiction well. There were subliminal tie-ins with cinematic depictions, too. The Fair's iconic architecture – the Trylon and Perisphere, a seven-hundred-foot-tall acute triangular structure accompanied by a spherical building and curved aerial walkway – was pure white. The design elements of the building, which contained one of two model cities of the future at the Fair (discussed below), are recognizably like the city of tomorrow in *Things to Come*. There was a shared message, too: wise technological innovations would shape a better social future.

The New York World's Fair was also good business. It marshalled advertising and covert promotions to preview a rosy corporate future. Not only science fiction fans were enthralled by the spectacles. Fair attendance, which exceeded 45 million, was bolstered by middle-class visitors seeking optimism at the tail-end of the Depression. Westinghouse centred an advertising campaign on the comic-book Middleton family, visiting the fair to see the consumer wonders of the near future. The Fair allied predictions with visions of the products of corporate technology and their related social benefits. Impressive corporate pavilions ranged alongside, and typically dominated, some thirty national pavilions and other exhibition spaces. General Electric, for example, demonstrated fluorescent lighting for homes, touted to be long-lasting and gentle on the eyes. Bell Labs showed off the *voder*, a voice synthesiser operated by a piano-like keyboard. International Business Machines (IBM) demonstrated its core information technology – the punched-card tabulating machine. Each invention would hasten a more convenient and empowered society.[28]

General Motors' Futurama exhibit was a grander vision of the technological future. It carried audiences around a three-dimensional view of an imagined city a mere twenty years ahead. Designed by architect Norman

Bel Geddes, the animated model covered some thirty-three hundred square metres and displayed streamline-style architecture, power-plants, multi-level expressways, and zoned urban activities (figure 5.3). Highlighting the intentionality of wise designers, the attraction was a paean to the techno-logical fix. In this rational city, buildings would be cost-effective and com-fortable, with ample conveniences, space, and sunlight. Industrial activities would be segregated from living and shopping districts. Traffic flow would be automated to rid city roads of congestion and to ensure comfortable pedestrian dominance of neighbourhoods. The exhibit showed scientifically managed agriculture, streamlined cars, and towering skyscrapers.

Some of these futuristic elements were available already, notably in some German cities where the unornamented lines of 1920s Bauhaus architecture vied with the smooth curves of Hitler's contemporary Autobahn projects. Even so, this large-scale thinking was closer to the scale of the Technocrats than to most contemporary American designers. The grandeur of thinking combined with the size and detail of the model overwhelmed audiences. Adnan Morshed suggests that, theatrically, the conveyor system that moved viewers' seats higher and lower over the model provided them with an om-nipotent view; by transforming audiences into seers, this may have subcon-sciously suggested human powers to reshape social life through technology. Bel Geddes himself described the exhibit as depicting an integrated redesign of modern society, "a new kind of civilization in which industry, finance and labor will all find greater employment – a vision of new frontiers of progress waiting to be conquered by those who will pioneer around the cor-ner of tomorrow." Futurama gained effusive reviews and proved the most popular attraction at the Fair.[29]

Although the 1939 New York World's Fair shaped contemporary American visions of the technological future, it lay midway along an established lineage of prior and subsequent international expositions. As summarized by Arthur Chandler, these fairs "set in motion one of the great rituals of Progress – the belief in the application of technology for the improvement of the quality of life." This acquired faith was closely allied with the notion of the technological fix – namely, that wise designers could effect positive societal benefits.[30]

As described earlier, technological optimism, conflation of technological and social progress, and implicit faith in fixes were not new, and World's

Figure 5.3 Forecasting the technologized city. Futurama exhibit, GM Pavilion, 1939 New York World's Fair.

Fairs played an important role in disseminating these beliefs. The model for such international events had been set by the Great Exhibition of the Works of Industry of All Nations, which took place in London's Hyde Park over six months in 1851. In turn, the Great Exhibition scaled up the ambitions of France, which had mounted eleven national expositions between 1798 and 1849. As the first international fair, the Great Exhibition provided a template for its successors. A dozen followed it (five more of them in France) by the end of the century. The trend in these expositions was to display material accomplishments of nations – new inventions, prototypes, and products – to showcase national industries, boost consumer demand, and encourage exports.[31]

Throughout the second half of the nineteenth century, most had exhibition buildings to show off the latest inventions, on the one hand, and cultural products such as art, music, and dance, on the other. The Columbian Exhibition in Chicago in 1893, however, introduced a more integrated vision of a utopian urban environment. The gleaming architecture of the "White City" represented an idealized Chicago of the future and included separate buildings dedicated to American manufacturing, electricity, and anthropological science. The mock-city scenario gave fairgoers a direct experience of the clean world of tomorrow and became an influential design guide for American architects and planners. Indeed, Bel Geddes's Futurama, fifty years later, was to depict just such a city, but one in which a new technology, the automobile, shaped social life.[32]

By 1933 there had been twenty world expositions following broadly similar formats. Collectively, they provided a liturgy for public expression of modernist faith. So popular were such expositions that participants founded the Bureau International des Expositions (BIE) to regulate these celebrations of industrial enthusiasm. Exhibitors ranged from national and regional governments to corporations, entrepreneurs, and even religious denominations. Like the notion of the technological fix itself, the Fairs hosted by the United States in 1893, 1933, 1939, 1962, and an unofficial "Universal and International Exhibition" in 1964, were the most starkly oriented towards the technological progress promised by American industry and its anticipated consumer benefits. Each of them promoted the familiar paired technological themes: the inexorable advance of invention and the social benefits carried in its wake. As Cheryl Ganz observes, the organizers of the 1933 "Century of Progress" exposition in Chicago believed that "progress rides on the swell of technological innovation" and had a deterministic vision of world progress through consumerism. Their views reflected the Fair's deterministic and depressing theme, "Science Finds, Industry Applies, Man Conforms."[33]

Foreign versions softened the theme. The motto of Expo 58, the Brussels World's Fair (Belgium, 1958) was "A World View: A New Humanism," meant to tie together the aims of world peace and social progress. It opened a year after the formation of the European Economic Community (EEC), forerunner of the European Community, which aimed to bring about the integration of European economies. According to the organizers, however, the key driver of economic and social progress was scientific analysis and engineering development. Atomic energy was expected to herald a new nuclear age and social benefits. The Fair's iconic building, the Atomium, modelled the unit cell of an iron crystal as a symbol of collective scientific progress but evoked the wonders of atomic power for most fairgoers. Expo 58 was the first public venue to display an operating nuclear reactor in the US pavilion and depictions of atomic-powered icebreakers in the Soviet Union pavilion – both opportunities to argue for the role of technologies designed to shape their respective societies. Their displays promised peacetime benefits for modern countries. This openness closely followed a thawing of nuclear secrecy, which had come exactly a decade after the atomic bombing of Japan. The 1955 International Conference on the Peaceful Uses of Atomic Energy, held in Switzerland, had revealed to the scientific participants a joint desire to share findings for civilian benefits worldwide. The new political openness trailed the Eisenhower administration's more cynically inspired "Atoms for Peace" initiative launched in 1953, which had sought to defuse the military arms race and its looming threat for the United States's nuclear ascendency.[34] Britain had triumphantly opened the first civilian nuclear power station in 1955 and, during the Fair, had a new experimental breeder reactor nearing completion. The exposition communicated a clear vision of the technologically liberated future for attendees but, like some previous fairs, was criticized for awkwardly juxtaposing this modernist vision with seemingly more primitive cultures, specifically a "living exhibit" of colonial village life in the Belgian Congo. The affair seemed to demonstrate the implicit technological faith of the participating countries and their exhibitors, suggesting that social values and cultural traditions would be subordinated in the brave new world of directed technological progress.[35]

The next World's Fair, the "Century 21" Exposition (Seattle, USA, 1962), picked up the pace of technological promotion. It was a product of Cold War politics and culture. Initially planned as a celebration of the expansion of the American northwest, the organizers realigned it as an exposition intended to demonstrate that the United States was ahead of the Soviet Union in technological advances. The subject had been a public and political concern since the 1957 launch of the first Sputnik satellite by the Soviet Union and the widely acknowledged and continuing lag of the American space program. John Kennedy's campaign claim of a "missile gap" between the United States and Russia in the 1960 American elections was argued to have been a factor in Eisenhower's narrow defeat.[36] Like the New York World's Fair on the eve of the Second World War, Century 21 portrayed American industry and its technologies as the source of the good life. Few European countries were represented beyond the Fair's "Boulevards of the World," a shopping area. The Soviet government declined to participate, and other countries in the Soviet sphere were not invited.[37]

It was, notes author Bill Cotter, "a Fair that would focus on the future." The "science and space focus" was in its title and in its motto, "Living in the Space Age." This orientation was a good fit not just for contemporary politics and enthusiasms but for business, too. Boeing, one of the world's largest manufacturers of aircraft and a contractor for the nascent American aerospace industry, was based in the city. Seattle councillors reasoned that the World's Fair would showcase industry and encourage inward investment. Economic aims created technological outcomes: months after the exposition closed, Boeing won the contract for the first stage of the Saturn V moon rocket and announced production of the new 727 jetliner.[38] Given the absence of the Soviet Union, the US exhibit dominated the World of Science pavilion. It focused on recent developments in genetics, atomic energy, and astronomy and showed off models of satellites and orbital capsules. The heart of the exposition, however, was "The World of Tomorrow." Visitors were transported in a "bubbleator" to witness the Seattle of tomorrow as well as an individual home. Visions of social groups and interactions were largely absent. Imagery contrasted the choice and individuality of technological conveniences with the regimented uniformity of new housing developments such as the Levittown suburbs of the 1950s. Wisely designed technologies, it suggested, would generate social benefits and values.[39]

Corporate visions of the technological fix were stage-centre. In the building were exhibits by Pan American World Airways, General Motors, and RCA, displaying the aircraft, cars, and media of the future. In a separate pavilion, "The World of Commerce and Industry," the Ford Motor Company had a spaceflight simulator and concept car of the future, and Standard Oil Company vaunted the history of petroleum fuel production; Bell Telephone provided its own film and exhibit extolling the conveniences of videophones, automatic dialling, and conference calls.

Technologies had visual impact and reached wide audiences and age groups. Images of the iconic Space Needle and Monorail were reproduced around the world and provided a template for representing how the social world of tomorrow would be deliberately shaped by technologies. *The Jetsons*, a cartoon spin-off of *The Flintstones*, depicted just such a world in weekly telecasts for children and their parents. The original broadcasts of the series coincided with the Fair, and it has been in syndication in the decades since. For teens, an Elvis Presley film, *It Happened at the World's Fair*, was launched to coincide with the exposition. Elements of the Fair's futuristic style, later known as *Googie* or *Populuxe* architecture, were reproduced in fast food restaurants, laundromats, and bowling alleys as emblematic of the future-oriented early 1960s, just as *Things to Come* and the 1939 Fair had represented the streamlined future through art deco. As for Seattle itself, the Fair provided architecture and transport that have not only remained in use but have continued to impress visitors and engender a remarkably irony-free appreciation.[40]

Consolidating this vision of the American technological future was another unofficial World's Fair in the United States two years later, held at the original 1939 site in Queens, New York. It took place as Alvin Weinberg was beginning to publicize his idea of the technological fix and, even more than earlier Fairs, it embodied technological confidence and faith in beneficent designers. Federal government participation, as with Seattle, focused on the display of space hardware, now updated to preview the Apollo program and "unmanned" probes. Communications and weather satellites impressed visitors with examples of how government-funded technology would soon transform their lives.

But the New York Fair marked a new scale of corporate promotion, too. Forty-five corporations took part as exhibitors, nearly all of them American. IBM displayed computer terminals, modems, and other apparatus; General Motors updated its Futurama exhibit, which again depicted a model city of the near future; and Ford offered a ride with a futuristic urban background. A less visible corporate dimension was the involvement of Walt Disney in creating exhibits and rides for the 1964 World's Fair. Many of them moved to Disneyland in Anaheim, California, when the exposition closed.[41]

Disney had a long-standing interest in technology. His original theme park, Disneyland, incorporated Tomorrowland as one of its nine attractions when it opened in 1955. The entertainment zone of Tomorrowland had rides simulating rockets, submarines, and cars of the future. Disney also produced television programs, animated short films, and children's books on space flight and nuclear energy. His networked television show, *Walt Disney's Disneyland*, included episodes devoted to spaceflight three years before NASA was inaugurated. The 1955 Disney television programs were reviewed in popular magazines such as *Popular Science*, leading to their rebroadcast on television and eventual release in theatres.

Disney's vision was not audience-led: it was the product of enthusiasts who appreciated the power of technologies to produce beneficial societal outcomes. He drew on advisers who had played a similar role in informing a series of articles about spaceflight for *Collier's* magazine in 1952. The ideas were neither home-grown nor drily scientific. Consultants for the programs included Wernher Von Braun (1912–1977), who had directed the Nazi rocket program during the Second World War and who had then headed the US Army rocket design program at Huntsville, Alabama. The principal Disney science consultant throughout the 1950s was instrument physicist Heinz Haber (1913–1990), who had been brought from Germany by the Americans after the war.[42] Haber wrote a popular Disney children's book and film short on nuclear energy for television and schools, and later built a career in Germany as a popularizer of science and technology.[43] A third German, Willy Ley (1906–1969), was an equally important consultant for Disney's technological enthusiasm. Trained in science and a participant in the German fad for rocketry during the 1920s, Ley had left Germany when Hitler came to power. He extended his career as a popular science and science fiction writer in the United States. The input of these three influencers resulted not only in the Disney television programs and children's books but also informed the Rocket to the Moon ride in Tomorrowland (figure 5.4). Such efforts seeded public optimism and expectation at a time when satellites did not yet exist and when there was little American policy interest beyond the development of military missiles. Yet this was not mere technological fixing: Disney's imagined future provided solutions for social yearnings yet undefined.[44]

This union of technophile and corporate interests proved durable. The Tomorrowland attractions were revamped in a subsequent stage show at the 1964 New York World's Fair, the Carousel of Progress, which thereafter became a repeatedly updated attraction at the Walt Disney World Resort in Florida. The attraction had been funded by General Electric to showcase its consumer products. In four acts, it recounts a Disneyfied history of technology since 1900 through the eyes of a typical American family, represented by animatronic robot characters. The revolving stage depicts the social effects of the introduction of electricity, faster transportation, and home appliances. The final act depicts the twenty-first century, featuring

Figure 5.4 Proselytizers of progress. Disney advisors Heinz Haber, Wernher
von Braun, and Willy Ley planning television program, July 1954.

voice-activated electronics and capped by a song, "There's a Great Big Beautiful Tomorrow." Punctuated with corny humour, the homespun narrative
depicts social progress and a comfortable continuity at the family scale.
The subtext communicates technological effects, rather than choices, and
the incremental improvements to be anticipated in a market-driven consumer society.[45]

   Disney found that his technological enthusiasms and promises of inventive benefits were well received by popular audiences and, consequently,
scaled them up in his second-generation theme park. He had been planning
an experimental planned community as combination of utopian experiment
and a new business development from the 1950s. He initially conceived
EPCOT, the Experimental Prototype Community of Tomorrow, as an opt-
in residential community pampered by technological conveniences. The
Epcot project had enrolled futurists and writers among its original designers.
Ray Bradbury and Buckminster Fuller, for example, were design consultants
for its Spaceship Earth attraction, described below. However, it evolved into
a technology demonstrator for American corporations and opened in 1982
as part of the new Disneyworld theme park in Florida. In either form, the

Epcot project nuanced the trope of technological optimism and inevitable futures: it argued that the social future could be actively and deliberately designed by wise technologists. Its core was closer to the ideals of Howard Scott's Technocrats.

Yet this theme was also being adopted by American corporations. Disney's Epcot was conceived as a product-placement theme park. Major attractions and entire buildings had corporate sponsors who branded the hardware and handouts. Exhibits featuring Apple, Monsanto, AT&T, Sega, and IBM have served as exemplars of the successful American corporation. Their products – and little beyond their products – shape visitors' understandings of technologies and their cultural appeals. Epcot's "Innoventions" museum of technology, for example, selects topics that vaunt company achievements. Thus "Future Cars" is sponsored by long-term World's Fair sponsor General Motors; "Colortopia," exploring colour, is sponsored by the paint manufacturer Glidden; Honeywell's exhibit extolling home conveniences is appropriately named "Comfortville." Even an interactive attraction, "Test the Limits Lab," play-acts the Underwriters Laboratories tests of consumer products. This sponsorship by contemporary companies provides a highly truncated family tree in which surviving businesses recount their product histories, leaving untold the much more common stories of failed companies and the social fallout of ill-considered products.[46] Historian Michael Smith suggests that Disney's narratives are stereotypes that misinform uncritical and accepting audiences. He notes "the absence of agency and causality":

> Decisions about design and function, social needs and social impact, are rarely glimpsed below the shiny, streamlined surface. When edited for display value, our view of technology achieves an autonomous, effortless quality only slightly more pronounced at Epcot than in the rest of American society.[47]

Much more long-lived than World's Fairs, the Disney narrative has influenced audiences continuously from the mid-twentieth century. Historian Amy Foster notes that four of Disney's theme parks (in California, Florida, Japan, and Hong Kong) still host Tomorrowland, and that Epcot is updated periodically to incorporate new commercially seeded dreams of the future. She suggests that Disney's "glorification of technology ... downplays the complexities of our future." Indeed, Walt Disney, with consultants drawn from the most optimistic engineers of his day, developed narratives of technological progress that influenced susceptible audiences unfamiliar with the perils and side effects of real-world innovation. Like the earlier American World's Fairs, Disney attractions and media combined technological enthusiasms with entertainment for young, optimistic, and middle-class audiences.[48]

Subsequent World's Fairs off American soil addressed themes that were less simplistically technology-and-consumer oriented. Arguably the last of the great optimistic World's Fairs was Expo 67 (Montreal, Canada). Themed as "Man and His World," Expo 67 highlighted technological advance but with greater sensitivity to social choices and implications.[49]

Its American pavilion, an immense geodesic dome by architect Buckminster Fuller, was a structure symbolic of technological advance. Fuller (1895–1983) was an iconoclast designer and futurist whose career intersected many of the names and events discussed in this book. He had been an unimpressive student at Harvard College and worked in a variety of manual labour jobs during the First World War but was able to fund later projects from an inheritance. Like Howard Scott, he had frequented Greenwich Village in the late 1920s, and he saw deliberate engineering design as a means of transforming the human world and improving quality of life. Like many within and beyond the Technical Alliance, he was concerned about waste in industrial processes and modern life. A polished self-publicist, Fuller became known from the 1930s for his plan for the "Dymaxion car" and "Dymaxion house" (the term combined "dynamic," "maximum," and "tension," mechanical engineering qualities meant to evoke efficiency and scientific principles). The Dymaxion car was a long, sausage-shaped three-wheeled vehicle, designed to allow maximum carrying capacity while being manoeuvrable and fuel-efficient. It was exhibited at the 1933 Chicago World's Fair, the Century of Progress, as an example of the car of the future. Fuller's concept of the Dymaxion house was of a prefabricated structure that made efficient use of building materials, water, and power resources. His architectural concepts appealed to the American military during and after the Second World War because they permitted strong structures that could be mass produced in aircraft factories. Fuller's promotion of the geodesic dome encouraged rapid-assembly temporary structures but proved equally popular at World's

Fairs and counterculture communes during the 1960s, the Spaceship Earth attraction (1982) at Disney's Epcot theme park, and the "biome" structures of the environmentally oriented Eden Project (2001) in Cornwall, England, making them emblematic of the wise technological future. Fuller's careful analysis and unconventional designs shared aims and approaches adopted on a societal scale by Richard Meier after the Second World War. Both sought inexpensive and simple solutions for disadvantaged populations and adopted increasingly large-scale perspectives based on systems theory. Fuller argued that his non-traditional configurations could liberate individuals and their environments. Like Meier, however, he was criticized for focusing on rational engineering while downplaying the importance of cultural norms in making technological choices.[50]

Reflecting social ideals at the peak of the Vietnam War, Canada's Expo 67 was more genuinely international than most previous expositions, with sixty-two participating countries. Other contenders for World's Fair status succeeded it but passed with less attention and influence. Expo 70 (Osaka, Japan) was the last in the series of closely spaced expositions, adopting the theme of "Progress and Harmony for Mankind." Expo 92 (Seville, Spain) focused on "The Era of Discovery"; Expo 2000 (Hanover, Germany) was on "Man, Nature and Discovery"; Expo 2010 (Shanghai, China) was dedicated to the theme of "Better City, Better Life"; and Expo 2015 (Milan, Italy) aimed for "Feeding the Planet, Energy for Life." The smaller-scale "specialized expos" classified by the BIE typically had low international participation and scarcely captured regional interest. They nevertheless conformed to the theme of technology as the provider of human comforts. Especially when hosted by the leaders in technological exports – Japan, Germany, and the United States – the exposition themes were aligned not just with invention but, increasingly, with living environments and nature. These visions were, however, individualistic and consumer-aligned, and evoked few of the societal and progressive themes of earlier expositions. Public taste for these cultural events had petered out. From the end of the 1960s, World's Fairs ceased to be showpieces for national accomplishments, ideological ascendancy, or, most important, visions of the positive technological future.[51]

One likely reason for this loss of public esteem was the evident role of international expositions in promoting the interests of big businesses. There is a conundrum in this. In an age of rising consumerism and international markets, there was nevertheless a decline in popular support of national industries and perhaps a growing suspicion of the societal dimensions of corporate activities. Popular audiences were ever more eager for quick fixes via technology but vaguely distrustful of the mechanisms and motivations behind the toys. It is notable that the more recent version of Disney's Carousel of Progress attraction, revamped after the loss of sponsorship by GE, incorporates an ironic postmodern take on narratives of progress. The final act of the show was recast as "The Future That Never Was," imagining a fantasy retro-future in the style of Jules Verne.[52]

Communicating Confidence: Media Visions

For audiences that could not travel to World's Fairs and theme parks, other media reinforced the appeal of scientific approaches and expectations of the technological future. Like fiction, these, too, hinted at diverse philosophical assumptions about how technologies could be marshalled for society. Broadcast radio was just such an example. Having spread rapidly from the early 1920s, for a single generation it became the dominant entertainment medium in North America. The evolution of radio networks allowed advertising and sponsorship by large firms for national broadcasts of regular programs. The National Broadcasting Company (NBC, 1926) was the first and largest, providing separate Red and Blue networks. The Red network became associated with popular programming and the Blue network with news and public affairs broadcasts. It was paralleled by the Columbia Broadcasting System (CBS, 1927) and, following anti-trust legislation in 1941, the American Broadcasting Company (ABC, 1943) succeeded the Blue network.[53]

Radio shows helped shape and extend popular understandings of the role of technology in modern society. A fraction of the programming communicated science, invention, and science fiction. The most significant and long-lived of the factual programs was *Adventures in Science*, broadcast on the CBS radio network from 1941 to 1959. Timed for after-school listening, the program aimed to encourage an interest in scientific research and engineering for high school students.

A not-for-profit science education organization, Science Service, was largely responsible for such media promotion. It had actively encouraged similar audiences via regular syndicated newspaper stories about science and engineering from its origin in 1921. Its first campaign had been the pop-ularization of experimental amateur radio through newspaper stories. Radio amateurism had spun off from professional activities during the First World War, when many operators and technicians had been trained in the use of communications equipment. With the availability of war-surplus compo-nents and the explosion of voice transmission experiments from the early 1920s, amateurs kept pace with commercial development and expanding government regulation. Upon the United States's entry into the war in 1941, Science Service presented the cbs radio series and organized Science Clubs of America and National Science Fairs for high school students.[54]

Popular magazines and government also nurtured these adolescent en-thusiasms for the new technology of radio as a means of directing societal progress. From the early 1920s, the Bureau of Standards drafted informa-tional pamphlets; the Department of Agriculture fostered Boys and Girls Radio Clubs for adolescents who would master radio and serve as informa-tion conduits between government departments and farmers; and Science Service disseminated the information through feeds to major newspapers and popular periodicals such as *Good Housekeeping, Harper's Magazine,* and *Popular Science Monthly.* Such initiatives aimed to employ radio as a lever for rural modernization and public education.[55]

Yet public engagement also evinced more diffuse themes linked weakly to technological social design. A much larger radio audience followed sci-ence fiction serials. *Buck Rogers* first appeared in comic strips and a science fiction novella in 1928.[56] It was serialized on radio from 1932 and in cinema (Universal Pictures) from 1939. Based on a character devised by Philip Fran-cis Nowlan (1888–1940), the adventure stories were set five hundred years in the future and had the technological trappings common to science fiction of the era. Such examples clearly created greater impact as visual media than as published texts. Comic strips and cinema displayed hardware of the future and the contexts within which it was used. Artist Dick Calkins (1894–1962) drew the *Buck Rogers* comic strip and wrote for the radio serial. The stories depicted clothing that was rationally close-fitting and synthetic, although styled and patterned into non-utilitarian fashions; vehicles and buildings were streamlined, following contemporary scientific findings about wind resistance and power consumption, although this had no logic in the vac-uum of space; electrical machines, control panels, and televisors hinted at the powers of technology to control and surveil but displayed no apparent social benefits. The characters and their technologies became the template for subsequent science fiction such as the *Flash Gordon* comic strip (1934) and movie serial (from 1936).[57]

Photo-magazines were more directly influential in communicating a vision of how technology and society were interlinked. While initially directed at radical European audiences, the sublimity of technology became an increasingly popular American theme throughout this era. Halftone printing had been developed during the late nineteenth century, and newspapers and periodicals were transformed throughout the First World War by monochrome photographs.[58] During the 1920s and 1930s, artists and photographers represented modern technologies in dramatic forms. Among the most influential was Alexander Rodchenko, a leader of the Soviet Constructivist movement. By exploring unusual perspectives, he defamiliarized the modern world and emphasized the scale and power of modern technologies. Viewed from extreme perspectives, radio masts, industrial machinery, and power plants were among his unsettling subjects. The complement of these representations of hardware was imagery of people marching and interacting with their modern environments in machine-like ways. In the United Sates, photographer Margaret Bourke-White was among the first to popularize this photographic style for Americans in her images of New York skyscrapers, Michigan factories, and Montana hydropower installations.[59]

The new field of photojournalism embraced these graphic depictions of technology. Magazines between the world wars were flooded with printed images, and cross-talk between them generated increasingly dramatic and breath-taking views of the Machine Age. Soviet magazines such as *Kino-Fot* (USSR, 1922) and *LEF* (USSR, 1923) were among the first, quickly followed by *Broom* (Italy, 1924), *Arbeiter-Illustrierte-Zeitung* (AIZ, Germany, 1924), and *Sovetskoe foto* (USSR, 1926). A growing number of topical magazines such as *Vu* (France, 1928), *Regards* (France, 1932), *Life* (USA, 1936), *Look* (USA, 1937) and *Picture Post* (UK, 1938) repeated and further popularized these stirring visual themes. What had begun as a movement in avant-garde art and a medium for propaganda grew, especially in the United States, to inform photoreportage, interior design, and popular culture. The medium carried the message that technology was breath-taking and empowering.[60]

This form of modernist photography proved equally effective for technical magazines, which began to adopt such imagery, feeding back the excitement of new technologies to engineers and scientists. Typical among them were the covers and advertisements in the MIT *Technology Review* from the early 1930s, *Discovery* magazine ("The Magazine of Scientific Progress") in postwar Britain, and *Scientific American* magazine, especially after its relaunch under new editors in 1948. As with World's Fairs, the depiction of wise technological design was appropriated by American companies. The imagery represented corporate sponsors and was directed at readers who were likely to be practising scientists, engineers, and technical managers. Technology companies such as Bell Labs, General Electric, Union Carbide, and Du Pont provided full-page advertisements. They typically relied on modernist photographs or paintings to represent science, innovation, and progress in abstract generality. Thus a 1950 Union Carbide advertisement shows a god-like finger welding a girder, and a similar giant hand pours from a test tube in a 1962 advertisement entitled "Science Helps Build a New India." In a successful series of such advertisements, the company strapline was "A Hand in Things to Come."[61] Similarly, Sperry Research (a printing company) symbolized its "200 research people with over 2000 years of experience" by a ghostly stylized microscope, chemical glassware, and sine waves. The implication of such advertisements and magazine covers suggested that science is a supra-human force directing humans. People, when shown at all, were likely to be disembodied hands and eyes, free of emotional taint. Rationality inspired abstract thought and appropriate technical solutions. The overall tone communicated to readers was of science and technology as powerful and transformative when in the hands of wise experts (figure 5.5).[62]

Photojournalism ceded place to television during the 1960s as the dominant medium for technological enthusiasms. In more direct ways, postwar television opened new channels for communicating positive visions of technology and the future. *Watch Mr. Wizard* was a popular NBC after-school television series between 1951 and 1965, portraying a science hobbyist who demonstrated his latest home experiments to visiting children. Within three years of its first airing, the show was telecast on some ninety affiliated stations, and a growing network of Mr. Wizard Science Clubs attracted primary

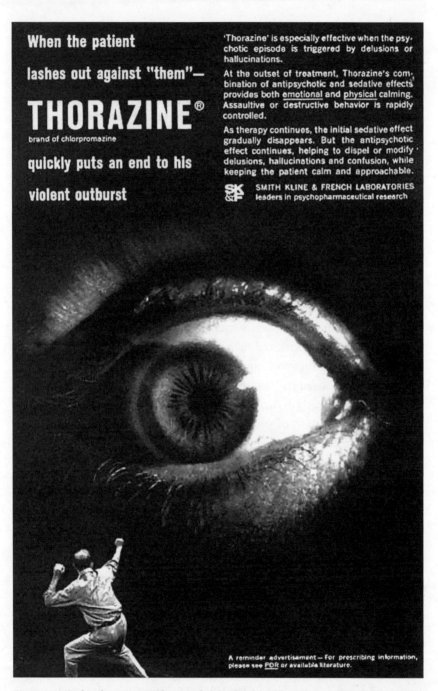

Figure 5.5 Technology as social control. (a) *Above* Smith, Kline, and French, 1962;
(b) *Opposite* Recordak, 1964 advertisements in *Scientific American* magazine.

The informer.

Putting a finger on a criminal suspect has become a lot easier in Baton Rouge since the MIRACODE® System by Kodak went on the police payroll.

Let a citizen outline six or seven characteristics of the felon who victimized him, press a couple of buttons, and there on the screen

flash the "mug shots" of *only* those criminals fitting the suspect's characteristics. Identification is quick and apprehension can be even quicker.

Funny thing is that the MIRACODE System wasn't designed for police work. Rather, it was designed to provide business with the fastest system ever devised to get information in or out of the files in seconds. And it works that way, thanks to microfilm.

It does seem a crime not to use it in your business, doesn't it?

For details, contact: Eastman Kodak Company, Business Systems Markets Division, Department PP-4, Rochester, New York 14650.

And ask about The Informer.

®Miracode is a trademark for equipment used in a coded input, retention, access and retrieval system designed and produced by Eastman Kodak Company.

RECORDAK Microfilm Systems by Kodak

---

school students across the United States and Canada. CBS provided more future-oriented technologies via *Walt Disney's Disneyland* show, as discussed above. In the deepening Cold War environment bracketed by the testing of the first Soviet atom bomb in 1949 and the launch of the first Soviet Sputnik in 1957, policy-makers sought to educate the public about the power and success of the United States's directed technological achievements, a goal actively pursued by media organizations.[63]

During the 1950s, artistic depictions of high technology as magazine art transformed science fiction fans into eager technological consumers. Newspaper comic strips and Disney's print publications of the period were to play a role in nurturing technological ardour with a growing corporate and patriotic tone. *Closer Than We Think!*, a single-panel cartoon image and explanatory text appearing in nationally syndicated Sunday newspapers from 1958 to 1963, reached circulations of some 19 million readers. Its creator, Arthur Radebaugh (1906–1974), was an illustrator and advertising artist who had worked principally in the American Midwest car industry. A competing comic, *Our New Age*, was syndicated to 121 American and foreign newspapers between the late 1950s and early 1970s. Its author, Athelstan Spilhaus (1911–1998), a South African geophysicist, sometime dean of the University of Minnesota's Institute of Technology, UNESCO ambassador, and board member of Science Service, recalled starting the comic because he "felt dejected about the state of the human spirit in the United States at that time in the late fifties." Spilhaus also chaired the scientific advisory committee of the American Association of Newspaper Publishers and was appointed by the Kennedy administration to direct the USA exhibit for the 1962 Seattle World's Fair. One widely reproduced anecdote alleges that the president said on meeting him, "The only science I ever learned was from your comic strip in the *Boston Globe*."[64] Both comic strips portrayed brightly coloured futures that Spilhaus and Radebaugh had encountered in their youth. The covers of the popular science and science fiction periodicals had filled American newsstands during the 1920s to 1940s: *Science and Invention*, *Popular Science*, *Popular Mechanics*, and, most dramatically, *Modern Mechanix and Inventions* (figure 5.6). These magazines blurred the line between science fantasy and consumer enthusiasm.

Since the 1930s, Radebaugh had been echoing these visions in his advertising art by depicting streamlined vehicles, futuristic cityscapes, and industrial robots for the Bohn Aluminum and Brass Corporation. He conjured vistas that included road networks for auto-pilot cars, robot agriculture, and polar cities under glass domes. Spilhaus, in turn, imagined computer-enhanced brains, education via automated remote teaching, and synthetic food as a worthy technological fix "to offset population increase." Some of his topics were somewhat more abstract than Radbaugh's consumer wonders; Spilhaus described his promotion of spaceflight, colonization of

Figure 5.6 Better lives through consumer technologies. *Mechanix Illustrated* (successor to *Modern Mechanix and Inventions*), 1957.

oceans, and other grandiose projects of the future as intending to encourage entrepreneurship and to reduce international tensions. In their distinct ways, the two vaunted the political power of American technical ingenuity. At the heart of these messages was a theme familiar to the Technocrats: deliberate technological problem-solving translated into social progress.[65]

During that decade, characterized by two American World's Fairs, the Apollo program, and, indeed, the rhetoric of Alvin Weinberg and technological fixes, television depictions of technology were just as upbeat and confident. *The Twenty-First Century* (CBS, 1967–69), presented by news anchorman Walter Cronkite, focused on the technology-rich future but hinted at mixed forecasts, too. Appropriately, the first episode explored science fiction as a predictive guide. The second season, however, focused on the wonders of computers as tools to educate the next generation of students – a classic technological fix – but tempered with the possibility of technological unemployment. The broader message was nevertheless that rational designers were hard at work designing and assessing technologies for human advance. "Inventing the future" became a catch phrase of this time.[66]

The technological optimism was apparent in the print media and broadcasts in other Western countries, too. In Germany, Heinz Haber, Disney's consultant for atomic energy and space flight, became a spokesperson for popular science and technology, presenting television programs such as *Professor Haber experimentiert*, and he launched the magazine *Bild der Wissenschaft*, which he ran from 1964 to 1990.[67]

The British mood, like that in the United States during the Cold War, combined optimism with a more overt degree of anxiety, although the economy was its focus. Labour prime minister Harold Wilson, elected in 1964, pinned the country's hopes on "the white heat of science and technology" as a source of jobs and the good life. His new minister of technology, Anthony Wedgewood (Tony) Benn (1925–2014), was prominent during the late 1960s in encouraging the adoption of new technologies by British industry. Benn promoted the forthcoming *Concorde* supersonic airliner and became a crusader for computer technology as a potential boon to society. His vocal advocacy of technological innovation was unparalleled in British politics and influential in shaping industrial and consumer trends.[68]

The simplistic optimism of technological fixes was more apparent in *Tomorrow's World*, launched in the summer of 1965 on BBC1, the more popularly oriented of the country's two national television services, as "a live, fun, science and technology programme." Its 1966 theme song was firmly deterministic about the technological future: "Tomorrow's world is coming / Whether we like it or not." More than 10 million viewers tuned in every week throughout the 1960s, looking for undemanding entertainment and, perhaps, confirmation of their expectations about the exciting consumer future. The program typically presented three or four stories per half-hour episode about new inventions and their likely impact. The unstated theme was that technology is an agent capable of solving human problems; this was a program that demonstrated hopeful technological fixes in multiple forms. Presenter William Woollard (1974–85) recalled:

> *Tomorrow's World* worked because it focused not so much on hard science, but on where science met technology. And because of that it attracted an audience that wasn't initially interested, perhaps, in science,

but *was* interested in the kinds of things that were happening ... it was focused on hardware ... A bunch of guys really enthusiastic about the science they were playing with, and really trying to make that interesting to a wide audience.[69]

The program was supplemented by an eponymous book that focused on the technological development of the oceans, spaceflight, robots, and synthetic foods. More disturbingly, its final chapter forecast a rational social world familiar to technology zealots over the previous century. "By the year 2120 AD," it advised light-heartedly, "every human being on Earth will live within a class structure based loosely on degrees of intelligence," governed by benevolent dictatorships in most countries, which would have "policies and aims ... virtually identical" to those of Britain. Underlining rational outcomes, the authorities would assign jobs, control holidays, and confine citizens in hospitals for compulsory treatment. Congenital illnesses would be eradicated by genetic engineering or government policies. Cultural traditions, democratic politics, and religious practices would wither away. The breezy tone of the text suggested that, despite the present-day unfamiliarity and distaste, modern societies would simply have to get used to the irresistible agency of technology and acquiesce to its direction.[70]

The casual magazine program, with its hidden assumptions, was not the only perspective on technology from British television, but it was the most popular over its thirty-eight-year run. By contrast, *Horizon*, a series broadcast a year earlier on the more seriously minded BBC2, continues as of 2020. The series was conceived as a platform for leading scientists and philosophers, and began with an hour-long documentary on Buckminster Fuller. Later episodes emphasizing the theme of leading-edge research presented alongside its social implications aimed to humanize the science. Topics typically were chosen to explore how scientific discovery influenced lives.[71] A more sociological tone, but still firmly oriented towards technologies and their consequences, was adopted by the Canadian series *Here Come the Seventies* (CTV, 1970–73). Over two dozen episodes explored topics such as the new genetics, mass transit, biochemical revolution, and pollution.

While such examples suggest a gradual nuancing of the understandings of anglophone publics, technological fixes remain attractive in popular culture beyond the United States. For example, a recent British program, *The*

Big Life Fix with Simon Reeve (BBC2, 2017–18), combines the notion of technological solutions with grassroots enthusiasms, as its presenter explains: "Our team's base is in East London. Known as a makerspace, it's one of a national network of inventors' hubs, crammed full of the latest technology. It's from here that our seven leading inventors will attempt to create fixes for people with nowhere else to turn."[72] The team's self-described "fixers" are identified as engineers, designers, and makers. While these categories blur the twentieth-century distinctions between professionals and amateurs, the program retains the separation between "experts" and "beneficiaries." The series seeks technological solutions for people with physical or other difficulties – for instance, to compensate the disabling hand tremors caused by Parkinson's disease via a counter-vibrating spoon. Other problems tackled by the team include designing a safety helmet for a young girl with brain damage "that will be less visible and help her fit in with her friends."[73] These laudable aims fit Alvin Weinberg's criteria of technological fixes: they provide engineering solutions to situations that might be addressed less quickly or simply, but perhaps more assuredly, by social, cultural, or other accommodations.

Selling Optimism: Business Bombast and Technological Faith

Literally promising the future, such television programs blur the distinction between neutral documentaries and promotional advertisements. The producers of *Tomorrow's World* relied on a stream of entrepreneurs eager to promote their product ideas to gain investors or immediate commercial exploitation. *The Big Life Fix* and its genre offer similarly identified problems and manufactured solutions, selected to appeal to audiences attuned to health and well-being. Media producers, engineering companies, and audiences themselves collude in shaping entertainment and profits. As a result, this confected programming may culturally reinforce shared beliefs rather than objectively assess the social value of technical innovations. In American culture, the "infomercial" acknowledges this combination of seemingly objective fact, entertainment, and product promotion as an overt sales tactic. It need not be intentional and cynical, though. Technological confidence has become a shared starting point for companies and consumers alike.

More general themes, akin to the *Scientific American* advertisements of the 1950s, are also offered by technology companies. The 2009 publicity campaign for the Apple iPhone 3, for example, was based on the slogan "there's an app for that."[74] Timed to promote the new App Store, the advertising promised software solutions for diet, health, and social engagement. Like Alvin Weinberg's promotion of technological fixes, the examples were pithy and discouraged careful evaluation. The original advertisements were simple and confident. Appreciating the advantages of technological fixes, they hinted, was the social marker of the new cool consumer:

If you want to check snow conditions on the mountain, there's an app for that. If you want to check how many calories are in your lunch, there's an app for *that*. And if you want to check where, *exactly*, you've parked your car, there's *even* an app for *that*! Yup, there's an app for just about anything.

Later ads in the series promoted apps for "fixing a wobbly shelf," reading an MRI scan or "a regular old book," communicating in Mandarin, and keeping in touch with friends (adding, with inadvertent threat, "you know what they're up to, where they are"). It reassured: "That's the iPhone: solving life's dilemmas."[75] Consumer aspiration and technological faith translated into social exclusivity, a truism long appreciated by manufacturers of automobiles and hi-fi equipment. Software developers innovated to supply solutions to users' perceived needs via their Apple, Android, and Microsoft software. For instance, Nextdoor, a private social networking app, was offered as the best way to keep in contact with neighbours – "real, verified members" – instead of via door-to-door chats or telephone conversations. EyeEm Selects promised to help select your best photograph for Instagram posts, and Pureple offered to plan outfits and organize closets. Others provide software to encourage mindfulness and adequate sleep. Several firms even offer apps to remind users to drink water regularly.[76] Extending the conflation of convenience and necessity further, the parallel development of the Internet of Things, in which hardware technologies communicate via the internet, allows householders (and potentially others) to monitor and control their mundane home appliances from their mobile device. While some of these developments invented solutions to hitherto unidentified problems, they also reinforced the belief that behaviours (even if only purchasing choices) could be shaped reliably by new technologies. And although the value of these developments for most buyers is at least questionable, it is notable that their genuine potential for empowering users with limited vision or mobility – notably via voice interfaces such as variants of the Amazon Echo (2014) and Google Assistant (2016), which can make such disablements socially invisible – have gained little market promotion or public recognition as egalitarian or socially enabling technologies.[77]

The range of possibilities for apps initially appeared exhilarating not just for consumers seeking personal empowerment but also for organizations pursuing social influence. The marketing slogan became the rallying cry of technology enthusiasts both in public and private organizations, particularly the information and healthcare industries. Instead of thirty-second television ads, these professionals published papers in their specialist journals. The majority uncritically trumpeted internet-mediated communications, sensing hardware and ubiquitous computing as unalloyed benefits to be adopted with all possible speed.[78]

A recurring theme of the promotions and their enthusiasts was the role of apps in enhancing or replacing existing social methods. Patients might be more conveniently monitored or diagnosed via phone apps, for example, than by queuing for busy clinic appointments. Phone sensors (accelerometer, gyroscope, pulse) might reliably record exercise and substitute for a fitness coach, and hikers might safely explore remote terrains without a guide, thanks to GPS and magnetometer sensors. Biologists could collect wildlife data from citizen-scientists equipped with phone cameras and SMS texting. Would-be musicians could tune their guitar, learn chords, follow a beat, and record their performances. Such apps altered hierarchies and drew new boundaries between experts and individuals. The novel possibilities transferred power in sometimes unappreciated ways; expertise flowed from traditional experts to programmers and, at times, to the app users but, at others, to associated companies that collated snippets of data about users to characterize them for additional commercial advantage.[79] Optimism about such apps encouraged further faith in technological determinism and in technological fixes for life's daily difficulties. Indeed, one analysis described "techno-spirituality" as a further area for growth, calling for apps designed to teach religious practices, provide guidance for daily life, and facilitate prayer exchanges.[80]

So pervasive was the Apple advertising campaign, and so effusive the converts' claims, that the motto was rapidly reflected back by critical users, subverted into sarcastic memes drifting through the internet. These might satirize smugly optimistic spokespersons, their unlikely solutions, or the brave new world that they heralded. A cartoon girl tells her dog, "I don't need you anymore. There's an app for that"; a small child using a tablet is captioned "Need to shut your kid up at dinner and you're a lazy parent? No problem, there's an app for that"; the image of a protesting teacher in front of a blackboard is captioned "There's an app for that! No, really, there is an app for that!" Yet despite ironic commentary from savvy millennials, the notion of software solving human problems appears to have been consolidated by, and within, modern business cultures.

Aiming at such corporate customers and public organizations, Dow Chemical Company launched a branding exercise for "solutionism" in 2012. Expressed as a legal trademark, slogans, and upbeat imagery, the company suggested that "solutionism" concerned the solving of human problems via clever technological solutions. The campaign hinted at confidence in technological fixes, with an added twist of empowerment for grassroots contributors (figure 5.7).

The compact term "solutionism" carried mixed messages, however. It had begun to appear in print from the 1950s and variously criticized confidence in policy, economics, or technology. Political scientist Samuel Huntington used it, for example, as a contrast to postwar conservatism in American international policy, which he characterized as "a critique of utopianism and "solutionism," and a respect for history and society as against progress and the individual." A manager-educator warned two years later, "beware of solutionism – the flabby optimism that there is a simple answer and that it will yield to the magic of a personality, brainstorming, sitting down and talking things over, or other tribal nostrums." You were guilty of solutionism, he noted, if you "launch into solutions before you understand the problem you are trying to solve." In a similar vein, a book on art philanthropy by Daniel Fox described the rejection of politics as technocratic solutionism: "Experts who practise solutionism insist that problems have technical solutions even if they are the result of conflicts about ideas, values and interests."[81] Two

Figure 5.7 Technological fixes as corporate slogan. "Talking together" aims to identify customer needs that company engineers can supply.

decades later, another writer criticized the elitism of such fixers: "Part of this hankering expresses a passion widely shared within this culture; for want of a better term, call it 'solutionism.' Solutionism is just a belief that for every problem there exists a solution; and successful persons are those who solve problems."[82] The term became much more popular in public discourse during the Reagan administration. Yet it still carried connotations of simple-minded government policy, as one Washington reporter observed:

> "Solutionism" is easy to spot. Its practitioners can be identified not only by their optimism, but by the use of certain key words and phrases. They like to talk about Marshall Plans and Summit Meetings, and call for the creation of new Cabinet Departments ... President Reagan, sometimes described as a conservative, is a solutionist par excellence. It is not easy for him to get through a campaign speech without telling an audience the right political choice can end their worries ...: "I don't think we need to waste any more time," he said, "listening to those who tell us what we can't do."[83]

Specifically *technological* fixes were promoted more optimistically during the Reagan presidency than during previous or later administrations. The 1984 "Strategic Defense Initiative" (SDI, known more popularly as the "Star Wars" program) famously aimed to end the arms race via a space-based defensive satellite network that would destroy intercontinental ballistic missiles. Like the science fiction fantasy that had thrilled cinema audiences seven years earlier, the imagined system would require powerful lasers and control systems that did not yet exist (and that were not, in fact, achieved).[84]

The president's broadcast announcing the deceptively faith-based program was received with dismay by technical experts who were assigned the task. It was straightforward even for laypeople to imagine simple countermeasures: reflective surfaces on Soviet ICBMs or crude and low-cost satellite interceptors. The program attracted growing criticism and eventually ridicule when its demonstrations of technological progress were shown to be cooperative targets conveniently vulnerable to massive, but still underpowered, ground-based lasers. The unalloyed optimism of the Initiative was also challenged by American scientists, engineers, and legislators across party lines. An analysis by the American Physical Society (the professional association of physicists in the United States) assessed the expensive program to be decades away from establishing whether such a system was even feasible.[85] To make matters worse, it was a technological fix in the most limiting sense. Beyond its purely technical inadequacies, SDI did not attempt to assess the likely economic, social, or political consequences of scaling-up the program to a working space-based system. The reaction to the program illustrates the rise of critiques by technical experts about Big Science and, to a lesser degree, Big Government (even if it did little to unsettle public confidence and media promotion of corporate rhetoric). The Star Wars program became emblematic of over-simplicity in policy analysis, naïveté in problem-solving, and unreasoning technological faith.

6 Tracing Life Histories of Technological Choices

Birth Notices and Obituaries

Faith in technological fixes relies on the expectation of positive outcomes and can even make shortcomings harder to see. This chapter consequently turns from promises to problems of technical solutions. It examines twentieth-century cases eventually identified as technological choices that had unanticipated negative consequences. As historian Thomas Parke Hughes concluded from such studies, "most technological fixes leave us in a fix."[1] And before the term had even been coined, H.G. Wells admonished:

> We have let consequence after consequence take us by surprise. Then we have tried our remedies belatedly. And exactly the same thing is happening in regard to every other improvement ... Isn't it plain that we ought to have not simply one or two Professors of Foresight but whole Faculties and Departments of Foresight doing all they can to anticipate and prepare for the consequences ...?[2]

Wells's warning has been repeated in new cases and disciplines (e.g., environmental law).[3] Yet these critical judgments are unequally subscribed to across expert disciplines and throughout technical cultures: views range from optimism about timely technological solutions to human problems, to outright dismissal of this putative societal tool.

Wider publics have been equally divided. We don't typically pay attention to the life history of technologies. It's demanding to keep track of how innovations begin and end, and the net implications of their existence. This kind of coldly objective and dispassionate gaze is unusual; each of us builds a personal relationship with the technologies we adopt over years or decades. As a result, it's more common to evaluate them retrospectively, as we would in writing the obituary of a relative. We might trace the positive aspects of that life and how it touched us individually, not dwelling on larger problems or controversies, and perhaps gently recalling a few imperfections and foibles.[4] But this rose-tinted cultural conditioning extends further when thinking about technologies instead of people. We expect succeeding generations of technology to be *better* than what came before. We anticipate improvement in design, trusting in a continual refinement of experience and wisdom. And if past technologies were quaintly imperfect, we unreflectively assure ourselves, the current ones are much better.[5]

Assessing the full lifecycle of a technology is just as unfamiliar to design engineers. Alvin Weinberg and his peers enthused about the potential applications of nuclear reactors but later acknowledged giving much less attention to reactor decommissioning and toxic waste.[6] The reasons for inattention can be guessed. A young designer is likely to be motivated by new possibilities more than by inherited problems. Re-engineering or cleaning up after designs of the past is seldom as exciting as addressing challenges of the future.

The goals of engineering projects are typically most malleable at the initial proposal and funding stages. The ambitions of the design will be defined for colleagues, clients, or customers, and its specifications carefully enumerated to ensure that agreed goals can be assessed. Experience with similar prior projects permits the designers to forecast likely technical difficulties and to allow financial estimation of associated costs. There is, however, a hidden side to development proposals. The identification of customer needs or beneficiaries may be a matter of convention. The intended audience for a new hydroelectric dam is likely to be a region or entire country, for example, but proponents can overlook the inhabitants of the valley it floods or the river ecosystem that it chokes off. If local inhabitants are compensated with new accommodation elsewhere, there may still be imponderable losses to their way of life. Such peripheral effects borne by others, calculable or not, are so-called externalities of financial accounting. Benefits enumerated in proposals are likely to be distinctive and measurable, while harms may be more ineffable and widely distributed, and consequently unassessed. Matters are worse for novel engineering projects. For the fresh approaches typically suggested as clever technological fixes, the promises can be seductive and side effects are unlikely to be well understood. Acute problems provoke quick, ill-considered fixes, aggravated by engineering enthusiasm and overconfidence.

This dominant tendency in engineering is contrasted by recent techniques such as "cradle-to-grave analysis" or "lifecycle assessment," which seek to assess the net effects of a technology.[7] A key feature is examination at multiple scales. The various steps or phases of an industrial process may be evaluated according to criteria that lie beyond conventional engineering training. In one variant, known as "ecobalance," attention focuses on environmental side effects generated throughout the lifecycle. Studying the production of cast metal products, for instance, might successively evaluate the mining of raw ore (and the environmental consequences of strip-mining) and smelting, casting, and fabrication processes (with production of contaminated air, water, and solid waste). Beyond the factory, the method would track distribution of the product, its usage (and possibly re-usage) by customers, eventual disposal, and, ideally, recycling of the component materials. The approach examines the impact of the product on numerous interacting systems.

But even this more integrated style of engineering attention is biased towards the material aspects of design. It pays particular attention to managing resources. Metals, for example, would be conserved in a well-designed closed-loop process of production, usage, and recovery; toxic by-products would be avoided or else contained and rendered safe; ecosystems would be sustained by avoiding the alteration of natural environments. This technical attention fits well with refined economics and environmental science. Yet the considerations are still limited to evaluating quantifiable aspects such as recycling efficiency, sustainability of fishing stocks, and overall cost.[8]

More ambitious versions of such lifecycle analysis refocus attention on sociotechnical systems. These are more problematic, though. Unlike engineering efficiency and costs, social and cultural considerations are difficult to quantify, and the logic of utilitarian trade-offs is less obvious. How significantly would a new high-speed rail line through a community harm or help it, for instance, and how can a sensible decision concerning it be made?[9]

As this suggests, the criteria of assessment are diverse and contentious. Evaluating technologies-in-the-round consequently demands a historical perspective for two reasons. First, an objective life history demands a complete life. The most insightful biographies typically draw on the subject's long and varied experience. Accounts of new technologies, like celebrity biographies, are notoriously difficult to assess while we are captivated by them.[10] Second, we need case studies that reveal the typical phases of this engagement: initial attraction, adoption, mutual accommodation, and sometimes rejection. By becoming familiar with the typical arc of technologies in society, it is easier to recognize characteristic features associated with faith in technological fixes.

Yet even for historians of technology, politics, or culture, careful scrutiny of a technology's score-card is uncommon. As discussed in the previous chapter, contemporary analysts write in an era of strong public faith in invention. Accounts of innovative technologies usually focus on their brief novelty and their positive social effects. To make matters worse, some of the best modern scholarship is founded on "microhistories" – careful accounts of the fine grain of a historical episode that may reveal complex relationships yet overlook the forest for the trees. There are exceptions, however. The rise of environmental history as a subdiscipline has encouraged an integrating perspective. Such accounts typically examine longer timescales, broader geographical coverage, or larger collections of social actors. A strong example of such coverage is the 2004 collection of technology history case studies edited by Lisa Rosner.[11] The best arguments against technological fixes generally highlight their unanticipated side effects and unintuitive outcomes. Careful analyses of fixes are seldom polemical, even when they reveal profound harms and recurring injustices. My central claim in this book is that technological solutions are widely trusted but that human instincts on the subject are inadequate.

This chapter consequently seeks to challenge cultural confidence in technological solutions by exploring cases of technological traumas. These selected examples cannot demonstrate general truths, but they are intended to suggest that unintended consequences are a common and under-recognized feature of engineering design. Prior studies have focused more frequently on the examination of industrial accidents and their technical and social causes.[12] The discussion here, however, emphasizes a longer chain of dependencies: infatuation with innovative solutions, gradual embedding of new technologies in societal systems, and the relatively slow and insidious effects caused by inattention to wider harms and longer time periods. The examples also suggest the comparable social, cultural, and legal lethargy when it comes to investigating and preventing the potential side effects of new technologies.

Biochemistry and Traumatic Outcomes

As explored in chapters 3 to 5, the two decades following the Second World War represented a high point for societal faith in technology, but the earliest popular condemnations emerged at that time and focused on the biological side effects of products of the chemical industries. These public expressions focused on specific cases of error. One of the most infamous was thalidomide, a drug developed in 1953 by Chemie Grünenthal and marketed in 1957 as a safe sedative medication. The drug was first reported to be the cause of profound birth defects some three years later by individual medical practitioners in Australia, West Germany, and Britain. The drug was withdrawn in 1961 in those markets but remained on sale during 1962 in Belgium, Brazil, Canada, Italy, and Japan. Worldwide, at least ten thousand babies were born with malformed limbs, ears, and internal organs, with nearly half dying in infancy. Seven members of the chemical company were brought to trial for inadequate testing, hindering withdrawal, and suppressing evidence. As was typical in the loose pharmaceutical licensing context of the time, thalidomide had been tested only on lab animals, not humans. More generally, it was not widely appreciated that chemicals ingested by a pregnant woman could selectively harm the development of her foetus.[13]

The tragedy fitted uncomfortably into the cultural attitudes of the period. Wonders of medical science – antibiotics, mobile X-ray units to detect tuberculosis, and the Salk polio vaccine – offered visible improvements in public health. For contemporary audiences, the thalidomide narrative segregated the case from such achievements, characterizing it as an aberration attributable to inadequate scientists or irresponsible companies rather than to societal over-confidence in goal-directed biochemical innovation.[14]

The case of DDT, on the other hand, has been credited for the birth of the environmental movement and an increasingly critical public from the 1960s. A Swiss chemist, Paul Hermann Müller, had discovered in 1939 that a chemical first synthesized over a half century earlier was an effective insecticide. He and colleagues at J.R. Geigy S.G. explored dichlorodiphenyl-trichloroethane (DDT) and its chemical variants, and performed field tests to investigate its effectiveness in eradicating flies, moths, and potato beetles. The compound was sold in neutral Switzerland for agricultural use from 1941 and, in another formulation for human usage, as an effective treatment for lice and fleas from 1942. Through its national branches, Swiss Geigy distributed samples to government laboratories of the major wartime powers. That year, the German government approved its application against plagues of potato beetles that threatened wartime food production. The chemical was also tested in Britain and in the United States (by the Office of Scientific Research and Development [OSRD] in Florida) to establish its toxicity for insects and safety for human subjects, many of whom were conscientious objectors. Encouragingly, DDT was found not to cause skin irritation, was free of smell and taste, and longer-lasting than known insecticides, remaining effective on the skin for up to a month.

The urgency of warfare encouraged quick technological fixes. Finding ranks decimated by cases of insect-borne diseases, the military had been combatting mosquitoes with campaigns to scorch and oil military environments to prevent larval production. They consequently chose the effective and seemingly innocuous insecticide as a better alternative. In 1943, DDT began to be deployed as a dusting powder and spray by Allied forces in the Mediterranean, Burma, and Pacific campaigns to control insect populations responsible for the transmission of malaria and typhus. Winston Churchill reportedly found the results astonishing, and the large-scale international deployment of DDT appeared to confirm its high toxicity for insects and safety for humans. So important was the benefit for soldiers and civilians that Paul Müller was awarded the Nobel Prize in Physiology/Medicine in 1948. The wartime perspective of applying new technologies to conquer enemies endured over the following decade: the chemical was central to the global Malaria Eradication Programme begun by the World Health Organization (WHO) in 1955.[15]

The same battlefield single-mindedness motivated the domestic uptake of DDT, which became available for civilian use in the United States immediately after the Second World War. Having been vetted by government labs, adopted wholesale by the armed forces, and promoted by government departments, DDT quickly became popular for American agriculture and households. The Department of Agriculture provided articles to farmers' periodicals and distributed pamphlets to homemakers, the primary audiences (figure 6.1). One counselled that *oil*-based DDT products might catch fire near an open flame and so should not be applied to livestock and pets but advised – perhaps with too many qualifications to be compelling – that "in the United States not a single case of DDT poisoning in humans has ever been proved when the material was used against insects." The pamphlet proceeded to instruct homemakers on how to rid dwellings of their insect enemies: roaches, black flies, clothes moths, carpet beetles, bedbugs, fleas, lice, silverfish, ants, ticks, wasps, and hornets.[16] With equal authority, *Science Newsletter*, distributed by the not-for-profit Science Service organization, touted DDT as a wonder chemical and fed stories to magazines such as *Better Homes and Gardens* and *Collier's Weekly*. *Time* magazine and *Popular Science* suggested that DDT would become the most important consumer benefit of the war, and, alongside the atom bomb, the chemical was lauded as a revolutionary scientific advance. Firms echoed the sober government messages more enthusiastically. Penn Salt Chemicals, manufacturer of Knox-Out DDT powder, published a June 1947 advertisement in *Life* magazine depicting a delighted farmwoman in a happy cartoon farmyard with vegetables and farm animals singing the motto, "DDT is good for me-e-e-e-!" Within a year of its introduction, companies were competing in a chemical arms race to offer cheaper, more potent, and more convenient forms of DDT insecticide. Consumers could choose products that included aerosols, powders, dusts, fog machines, residual spray for walls, DDT-impregnated wallpaper, and polyurethane coatings.[17]

The miracle compound continued to be investigated, however, to confirm its lack of potential human harms. Industrial production of synthetic chemicals had periodically raised concerns when correlated with workers' health problems. Occupational illnesses often become evident before effects of

Figure 6.1 Confident chemistry: DDT promotion to rural households.
(a) US Department of Agriculture, 1947; (b) *Nebraska Farmer*, 1945.

products are noted in the general population because of the higher exposure of employees in the industry. Action, typically in the form of legislation concerning occupational health to limit workplace exposure or to withdraw products, may lag far behind. For example, the discovery of radioactive materials at the turn of the century had spawned firms to exploit their reputed health-giving properties (figure 6.2). While spreading these beliefs in popular culture, companies became aware of occupational dangers. As early as the 1920s, bone cancers had been correlated with the ingestion of radium that occurred when the painters of luminous watch-dials licked their brushes. Newspapers of the day reported the issue, and a small number of the workers won legal compensation a decade later, but radioactive waste from the contaminated sites was not completely cleaned up until the 1990s.[18]

Along the same lines, coal-based synthetic dyes, correlated with bladder cancer from the 1890s, were investigated by Wilhelm Hueper forty years later with funding from Du Pont de Nemours chemical company, which had become the principal American producer of synthetic dyes during the First World War. Hueper (1894–1978) deduced, to the chagrin of his funder, that bladder cancers are highly correlated with the chemicals and typically begin

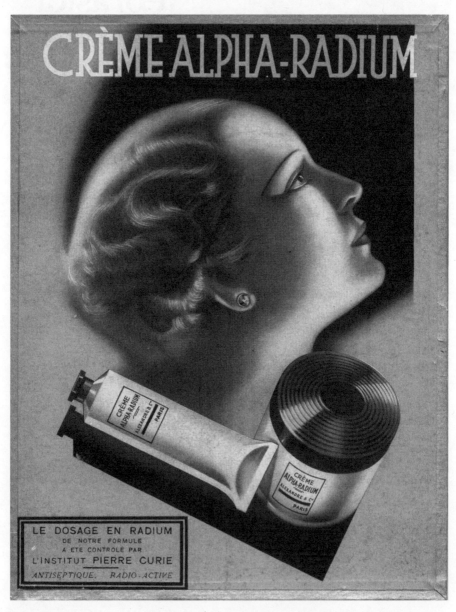

Figure 6.2 Scientific beauty. Youth-regenerating cosmetic advertised as "antiseptic and radio-active," 1933.

to manifest in dye production workers some thirty years after factories open.[19] A similar refrain of reported harms, slowly confirmed scientific evidence, and corporate sluggishness surrounds asbestosis, the scarring of lung tissue by breathing asbestos dust. First noted by physicians during the 1920s, it was widely acknowledged and acted upon only during the 1960s. In the United States, the Occupational Safety and Health Administration (OSHA) was founded in 1971 to deal with such, by then, familiar cases of side effects and resistance from the chemical and manufacturing industries.[20]

Geigy itself funded a 1944 toxicology study at the University of Cincinnati (sited near its American DDT production plant); it revealed that DDT, when supplied in large dosages, tended to accumulate in fatty tissues and milk in mammals. A year later, the *Journal of Wildlife Management* devoted a special issue to DDT and wildlife, and the US Fish and Wildlife Service cautioned against widespread usage owing to reports about the chemical's unintended effects on the mortality of fish and birds.[21]

Despite these early clues, the leap from studies of human toxicity to side effects in other species did not occur until the publication in 1962 of Rachel Carson's book, *Silent Spring*. Carson (1907–1964) was a marine biologist who had spent her early career at the US Fish and Wildlife Service while gradually reorienting her work towards popular science and nature writing for radio, magazines, newspapers, and books. Throughout the 1950s, she became best known for her second and third books on the life history of the sea, the earlier of which was serialized in magazines such as the *New Yorker*, *Science Digest*, *Reader's Digest*, and *Yale Review*.[22] As her interests in wildlife conservation grew, Carson's writing projects culminated in *Silent Spring*. Unlike her previous publications, the book focused on environmental harms caused by industrially manufactured chemical products and, specifically, pesticides. She cited Wilhelm Hueper as an inspiration for her wider focus on the environmental side effects of industrial chemistry.[23]

Carson collated evidence that DDT tended to accumulate in mammals in treated environments, killing outright at high enough concentrations. Like Hueper, she found evidence that DDT was a carcinogen and a trigger for human illnesses. At lower accumulations, the chemical specifically affected the thickness of eggshells of birds, making successful breeding in the wild unlikely due to fragile eggs. She also underlined that use of such pesticides causes rebounding pest populations as subsequent generations are spawned by surviving members that have greater resistance to the chemicals.[24] More contentiously, she argued that the chemical industry misinformed government bodies and resisted external scrutiny. Translated into the language of technological fixes, DDT was a short-term and unsustainable solution, causing more harm than good. Carson recommended that pesticides should be used sparingly and that a biotic approach (employing parasites and predator species) and ecosystem perspective should replace reliance on such biochemical fixes.

Carson's challenge to modern technical solutions coincided with American popular awareness of the thalidomide tragedy unfolding in other countries. As with her earlier writing, chapters of *Silent Spring* were recast as instalments in periodicals such as the *New Yorker* and in *Audubon Magazine*, an important channel for amateur naturalists and conservationists. These previews triggered a backlash from chemical industry lobbyists, who published counter-statements, threatening legal action against the publishers and withdrawal of advertising from magazines that had serialized the book. Resistance from American Cyanimid, Du Pont de Nemours, and the Velsicol Chemical Company was rebuffed. In the year between the publication of the book and her premature death from cancer, Carson was widely acclaimed by academics and government bodies. Her criticism of the chemical industry was not unique: less successful publications of the period included *Our Synthetic Environment*, published a few months earlier by anarchist philosopher Murray Bookchin, who later more generally critiqued modern technological faith (as discussed in chapter 7).[25] Despite challenging prevailing technological industries and going against the grain of cultural faith in modernity, Carson's careful scientific evidence was supported by peers. Her clear and poetic style made the book compelling for large audiences.[26]

Lifestyle Fixes and Failures

Alvin Weinberg had admired Carson's work, which sensitized him to the direct health consequences of human-made chemical products in his own field, such as the radioactive by-products of nuclear reactors. Yet his promotion of technological fixes just a couple of years after her death tended to fall short of any cautionary stance about such narrowly conceived solutions. His off-the-cuff example of cigarette filters, for example, considered a simply defined problem (lung cancer, strongly correlated with smoking) and a social context deemed unchangeable (smoking, smokers, and a national economy to which the tobacco industry made a sizeable contribution) to argue for an easily implemented technological fix: a cigarette filter to absorb the tar by-products then argued to be the primary cause of the disease. For Weinberg and his audiences, this undoubtedly looked like a good start: a straightforward solution that, while not perfect, might be supplemented by other, better, technological solutions or gradual cultural change.[27]

But cigarette filters were a simplistic solution to a scarcely analyzed problem. Designed to absorb relatively large particulates, they failed to remove dozens of carcinogenic gases. Filters tended to restrict airflow and nicotine in each puff, causing some users to smoke more cigarettes and to breathe more deeply, depositing smaller particulates deeper in the lung where cancers were more likely to begin. Studies of cancer rates for filter-tip smokers through the 1970s and beyond showed evidence of varying levels of reduction, none higher than 50 percent. And since the 1980s, passive absorption of (filtered) tobacco smoke by non-smokers has been recognized as a significant additional cause of cancers and has been correlated with chronic illnesses such as heart disease and emphysema.[28]

Such health implications have been pieced together laboriously by researchers at university labs and government departments, and hint at wider societal fallout from smoking with or without filter tips. The net economic implications have been similarly difficult to ascertain. The benefit of cigarette filters in sustaining corporate profits is a well-recognized dimension of their positive economic effects, but there are also negative externalities produced by the tobacco industry. Cigarette butts, for instance, pollute city streets and waterways, endangering aquatic life. They increase the cost of urban cleaning services and are responsible for accidental fires that increase costs for civic fire departments and hospitals. The economic costs of mortality from cigarettes are typically borne by consumers and the state rather than by the tobacco companies themselves.[29]

Cigarette filters did allow the tobacco industry to innovate in ways that expanded its customer base, creating more new smokers to consume more tobacco. Firms introduced filter tips, and advertising to support them, from the 1930s as a solution to throat irritation (figure 6.3). When the link between lung cancer and smoking was officially acknowledged by the US Surgeon

Figure 6.3 Rational solutions for smokers. Cigarette filter
advertising 1949–50, used with minor variations between
the mid-1930s and 1960s.

General in 1964, companies innovated further to maintain markets. Advertising praised the technical advantages of cellulose acetate, activated char-
coal, or micro-pores. Novelties included new filter-based smoking products,
designed to impart flavour or colour to the tobacco smoke and sometimes
aimed specifically at female or younger smokers. Combined with new tobacco formulations and porous surfaces that mixed the smoke with air, these

were marketed as "light" cigarettes, although some chemical components in the smoke were actually increased.

Thus, the case of cigarette filters belied Weinberg's optimistic expectations. Manufacturers' claims about the technological fix for health provided by filter tips did not translate into a reduction in lung cancer rates but, instead, actively delayed and worked against effective legislative and health policy measures. These technology-sustained consumption trends have gradually been reversed by a raft of traditional social measures introduced incrementally: legislation to restrict sales of cigarettes to minors (enacted as early as the 1890s in some North American political jurisdictions); regulation, and eventually prohibition, of tobacco advertising; public health education campaigns; targeted taxation; prohibition of tobacco company sponsorship of sports events; mandatory health warnings on cigarette package labels; prohibition of cigarette dispensing machines accessible to children; and, since the beginning of the twenty-first century, prohibition of smoking in a broadening range of public areas and jurisdictions.[30]

The case suggests that misplaced confidence in technological solutions, especially when promoted by powerful social interests, may act against alternative forms of problem-solving. As early as 1973, general studies of the shortcomings of technological shortcuts were focusing on reputed technical solutions to ill health. Sociologists Amitai Etzioni and Richard Remp, working for a philanthropic funder of social research, examined the wider implications of a half-dozen technological fixes, including the employment of the chemical disulfiram ("Antabuse") as a treatment for alcoholism. Their aim was to assess Weinberg's claims about technological fixes by evaluating whether this medical technology could solve problems of addiction "without treating the personality or requiring its restructuring." The authors identified the saving in time and cost – the replacement or reduction of expert labour – to be the key criterion in evaluating the utility of the medication. But, as they noted at the outset: "Almost never is there a clear answer [about whether] a technology 'works' or not, and the decision [whether] it works is much affected by the societal context in which knowledge is produced, communicated and used."[31]

Prescribed for alcoholism from the 1950s, disulfiram is a "counterdrug" that induces nausea and other intense hangover-like effects immediately after drinking alcohol. While the drug can be highly effective in ensuring abstinence from alcohol for days after it is administered (making it as effective as an expensive hospital stay), long-term side effects for the patient may include more extreme or long-lasting physical and psychological effects than intended. In the earliest study, nearly half the patients ceased attending clinics. Among the remainder, the proportion judged to have "recovered" from alcoholism was comparable to those who had taken other forms of treatment, such as conditioned reflex therapy and meetings of Alcoholics Anonymous. The original study noted that the Antabuse treatment required control of a panoply of social conditions, including a carefully managed clinic "experience," group patient discussions, support networks, and an energetic follow-up program to encourage continued sobriety. Considered in economic terms alone, Antabuse therapy thus had uncertain benefits. Its value was highly contingent on the context of patients' circumstances, healthcare funding, and support by ancillary services. Antabuse therapy was therefore not the clean-cut technological fix that Etzioni and Kemp had initially hoped to identify but merely one component in a complex sociotechnical system.[32]

Car Culture and Unintended Consequences

The advantages of automobiles are globally familiar after more than a century of enculturation. Yet the social problems that they originally addressed – the original technological fixes represented by the automobile – were seldom appraised and largely forgotten; the new problems they created have been appreciated only in the past two generations. Car dependence brought social side effects that were unplanned and insidious.

Cleverly innovated as a wide variety of intriguingly small motorized vehicles from the 1890s, cars had been taken up by affluent and adventurous consumers as transportation that carried a social cachet. Within twenty years of their invention, private automobiles were part of a complex ecology of urban transport. Public transport in the form of horse-drawn trams (running on steel tracks) and rein-steered omnibuses had been growing in most Western cities from the mid-nineteenth century. Trams (streetcars, in Howard Scott's interwar North America) were an urban version of the passenger train; omnibuses had evolved from earlier small horse-drawn coaches. Following increasingly regular schedules and expanding routes, both forms of vehicle carried factory workers for their shifts. Powered by electric motors from the 1880s, trams became the principal form of urban transport in modern cities throughout the first half of the twentieth century. Electric trams and hydrocarbon-fuelled heavy goods vehicles overtook horse-drawn vehicles after the First World War. In their time, they amounted to a technological fix for the widely recognized problems of noise from clattering hooves on cobbled streets, public health concerns about horse dung, and the rising death toll of overworked urban horses.[33]

Other transport inventions vied for attention. A craze for bicycles had peaked in the 1890s and was a more obvious culturally driven shift towards technology. Manufacturers had promoted bicycle riding as affordable transport for working men, as a technological means of liberation for young women now able to explore the countryside in unchaperoned groups, for private excursions by young couples, and as healthful exercise for all. Fashion followed social desire. Adoption by women of "safety bicycles" (the diamond frame shape and equal-sized wheels familiar throughout the twentieth century) was made more culturally acceptable by the invention of bloomers, or divided skirts. With the invention of small internal combustion engines, bicycles were converted into personal motor vehicles, a transition accelerated during the First World War. As with cars, using bicycles and motorcycles was preferable to walking and gave independence from public transport schedules.[34]

Private automobiles quickly propagated within this evolving system. Car designs can be understood in the language of technological fixes. They provided unique social advantages for their affluent owners, taking the place of carriages but without the burden of stabling a horse. Unlike trams and buses, cars were privately owned and typically unshared with strangers. By the 1920s, most were enclosed – segregating the passengers further – and benefitted from improved suspensions to protect passengers from adverse weather and jarring roads. Robust designs, such as the Ford Model T, improved mobility and promoted exploration beyond the city. Cars enabled convenient, unscheduled trips to suit owners' whims; they granted the privilege of independence and autonomy. None of these features suggested problems for the intended users. Indeed, social exclusivity and the freedom to explore were features commonly recognized by wide audiences; the perceived disadvantages of public transport were strong motivations for aspiring car-buyers. Ownership grew steadily in North America and at a slower but rising pace in European countries, especially after the Second World

War. An automobile-based life has been aspirational in most countries, with China in the early 2000s managing a rapid and large-scale transition from bicycle-based transport.[35]

But although railways had reached a peak of density and efficiency by the end of the First World War, road networks still catered to horse-drawn traffic. This was the first broadly appreciated non-technical problem and political challenge presented by automobiles. Unlike railways, privately financed but typically extended with encouraging government subsidies, roads were funded more overtly by local government and public taxation. The beneficiaries were relatively privileged car owners and the burgeoning car industry. There were additional public costs, too: a new bureaucracy, with new technical expertise and responsibilities, was grafted onto government. New drivers had to be tested and issued licences; private vehicles had to be inspected for safety and issued with permits; new legislation for appropriate road usage had to be drafted; and bad drivers had to be policed, judged, and fined by an expanded legal system.[36]

Automobiles were the core of a stable sociotechnical system within a half-century of their invention. Proliferating small firms had grown from, or consolidated, earlier manufacturing traditions and industries. Carriage-makers, bicycle manufacturers, and smithies developed growing sidelines, and new trades replaced the old. New firms grew in their wake to support cars. Petroleum-based fuels became dominant and were supplied from pharmacies, corner stores, and, by the 1920s, a growing network of service stations that could also repair and maintain the vehicles. Subsidiary manufacturers became established to supply spare parts and accessories such as rubber tires, (optional) headlights, engine oil, and spark plugs. New services also catered for newly mobile consumers and became integral features of car culture. Fast food outlets and drive-in movie theatres were common by the 1930s and spread along inter-city routes. Car-oriented businesses consequently began to stretch and merge city limits into urban sprawl. Technological momentum also made cities more car-centred. The production of millions of car engines, for example, provided momentum for trucks and buses sharing engine technology. After the Second World War, car manufacturers increasingly lobbied local governments to replace electric trams with diesel- and petrol-fuelled buses. Civic governments were often content to do so as it reduced their costs for laying track and electric lines to rapidly expanding districts at a time when growing petroleum refineries had realized economies in fuel production.[37]

As car manufacturers improved the reliability of their products, networks of corporate filling stations became implanted in cities and on roads as motorists' oases, and planners largely succeeded in establishing regional motor vehicle networks. Funded at regional and national levels, these aimed to minimize intersections, reduce diversions, and meet a minimal standard for carrying heavy traffic. Relatively short routes were trialled in the Long Island Motor Parkway in 1911, the Italian *autostrada* in 1924, and the Bonn-Cologne *autobahn* in 1929. A generation later, the Eisenhower administration approved the Interstate Highway System to consolidate the evolving American network, which incorporated growing sections of "freeways." The similar "motorway" system in the United Kingdom was begun with the first section of the M1 opening in 1959. Such routes, limiting access to high-speed vehicles, contributed to the growing economic momentum of trucking and the decline of rail freight in those countries. The net effect of this sociotechnical transition was the growing dependence of society on the standardized technology of motor vehicles and the decline of viable alternatives.

While these changes were seen largely as positive or at least as inevitable, a hint of problems began to be recognized after the Second World War. A serious case of airborne pollution – first described as "smog" (smoke/fog) earlier in the century to describe conditions in London caused by large-scale coal burning – was reported in the industrial town of Donora, Pennsylvania. The week-long event was implicated in the deaths of twenty persons, the illness of several thousand more, and increased mortality over the following decade. The Donora incident was attributed to hydrogen fluoride and sulphur dioxide emissions from the town's zinc- and steel-manufacturing plants, exacerbated by a temperature inversion that kept the airborne chemicals at ground level. The zinc-smelting plant, in particular, was an early suspect owing to the absence of vegetation for hundreds of meters around the plant. The factories continued operations over the following four days and, after a rainfall that altered the weather pattern, resumed operation after a single day's closure.[38]

While such industrial pollution incidents had been lamentably common over previous decades, they were typically associated with acute events following occupational exposure to toxic chemicals. A comparable association between car engines, airborne pollution, and environmental harm was not made until the early 1950s. Smog days became more frequent around Los Angeles, the American city with the highest usage of cars since the final years of the war. Analysis revealed that the photochemical smog was the product of sunlight, nitrogen oxides, and hydrocarbon compounds from car exhausts and petroleum refineries. The resulting chemical cocktail includes ozone, acid rain, and particulates. Initially, the problems were summarized as a "decrease in visibility, crop damage, eye irritation, objectionable odor, and rubber deterioration." These appeared to be smaller-impact versions of the Donora incident, causing mild and temporary nuisance. However, later medical studies revealed the chronic effects of low-level smog: correlations with respiratory diseases, cancers, low birth weights, and birth defects. Over the following decades, more dramatic episodes of smog were recognized in other modern industrial cities, particularly those with warm, dry climates and a large population of car drivers.[39]

A distinct form of air pollution fully attributable to petroleum-fuelled engines is tetraethyl lead (TEL). Manufacturers intentionally added the compound to fuel in 1921 as an overt and well-publicized technological fix after the General Motors research laboratory showed its effectiveness in improving engine performance. Although the dangers of lead poisoning had been recognized during the late nineteenth century from sources such as house paint and wallpapers, the only initial concerns about leaded fuels concerned incidents of accidental toxic exposure involving petroleum refinery workers. Concerns about airborne lead exposure from car exhausts developed only after the Second World War, when evidence grew for a historic rise in environmental lead contamination coincident with the introduction of TEL. Subsequent medical studies correlated tetraethyl lead in the environment with toxicity and, specifically, with chronic effects on children's school performance, even suggesting a correlation with the incidence of violent crime.[40] In a sense, then, calls for *removal* of lead from fuel constituted an "anti-technological fix" designed to yield social benefits – namely, healthy, well-educated, and law-abiding citizens.[41]

The most recently identified effect of internal combustion engines is their contribution to anthropogenic climate change. As a technological side effect, this is particularly elusive, affecting global rather than local environments and discernable over years rather than days. While pollution from car engines has increasingly been recognized, solutions to the growing problem were tardy and partial. As with other cases of technology-induced problems, the responses included corporate legal challenges, optimistic technological fixes, and legislation driven by societal concerns. The technological dimensions of solving the lead problem were a relatively minor component of its ultimate achievement. Engine manufacturers introduced tweaks to maintain performance and altered the materials used in piston valves that had been reliant on the lubrication provided by TEL. Petroleum refineries introduced additional additives, assured to be free of the toxic effects of lead, to aid performance and to reduce corrosion and fouling.

Much more influential was legislation. Responding to growing environmental awareness by citizens, the Nixon administration founded the US Environmental Protection Agency (EPA) in 1970. Three years later, the organization planned a phasing out of lead in petrol, but the Ethyl Corporation challenged the validity of scientific studies of environmental lead and its toxicity and opposed the EPA regulations in the federal courts. After a lengthy appeal process, the national scaling down of leaded fuels began in 1976, more than a half-century after their introduction. The petroleum industry negotiated delays in the process through the mid-1980s, ostensibly to assist the trucking industry and older cars for which conversion to unleaded fuels would be costly and impracticable. "Lead credits" allowed some fuel users to gain access to higher-lead fuels while encouraging an overall national reduction; service stations continued to offer leaded fuel alongside unleaded alternatives. By the 1990s, however, leaded fuels had been almost entirely replaced, accompanied by significant reductions in biological lead levels in the American public.[42]

In the same way, the problem of smog from engine emissions has generated technical innovations and consumer behaviours driven by legislation. Government action and environmental standards have been highest in California – arguably the American region most affected by smog – but have been followed by other American states and other countries. The Los Angeles County Air Pollution Control District was formed in 1947, initially with the remit of monitoring and controlling factory emissions. At state level, the California Department of Public Health formed the Bureau of Air Sanitation, which established the first American standards for air quality and exhaust emissions in 1966. The original standards considered hydrocarbons and carbon monoxide; additional compounds regulated later were nitrogen oxides (1971) and particulates from diesel engine fuels (1982). The national standards adopted by the EPA have been paler imitations of the Californian model. The state also sponsors research into low-emission engine technologies; in effect, the government funds innovation to address problems that were created by private industry a century earlier.[43]

The technological fixes for emissions have been varied but highly dependent on enforcement legislation and encouraging tax incentives. Catalytic converters were the earliest and easiest: aiming to allow continued use of petroleum-based fuels and sustain the petrochemical economy, they could be retrofitted to existing cars (although leaded fuels "poisoned" the catalyst, rendering it ineffective). Catalytic converters transform carbon monoxide and unburned hydrocarbons into carbon dioxide and water. Later designs also reduced exhaust concentrations of nitrogen dioxides. Such conversions of toxic pollutants into greenhouse gases exchange an acute problem (smog) for a chronic one (climate change). They also have associated problems, notably fires caused by contact with the catalyzer surface heated by the chemical reaction. Nevertheless, car buyers baulked at the additional cost of a few hundred dollars, and their uptake by manufacturers had to be mandated by EPA regulations for car models from 1975 onwards. The ultimate disposal or recycling of converters was little considered by manufacturers, making them a typically short-sighted technological fix.[44]

Electric cars were conceived as a wider-ranging technological fix. By avoiding petroleum fuels, they could also avoid noxious emissions. More important, electric vehicles promised to make car-owners independent of oil, for which the price had been quadrupled and coupled with a scaling-down of production in late 1973 by the Organization for the Petroleum Exporting Countries (OPEC). US domestic oil supplies had been supplemented by Persian Gulf suppliers for several years. The embargo was an economic shock to government and consumers alike, leading rapidly to energy shortages for industry and fuel rationing at filling stations. It also encouraged searches for alternatives to the petroleum-based economy of which internal combustion vehicles were a key component.[45]

Electric cars were not new: varieties had proliferated at the turn of the century. Before the First World War, one manufacturer sold several hundred per year, but sales fell after the launch of the much lower-priced Ford Model

T. Electrics had been marketed especially towards female drivers, whom manufacturers imagined requiring simple-to-operate vehicles for short trips from home. Even more than their hydrocarbon-fuelled competitors, electric cars of the early twentieth century were marketed as technological fixes for modern social problems (figure 6.4).

Heavy batteries and limited range were technical limitations, but socio-economic factors also limited the re-entry of electric vehicles after the oil crisis. A support network had to be built from scratch, while simultaneously competing with a well-entrenched car economy. Charging facilities, like filling stations selling petroleum fuels in the 1920s, were rare, making long trips risky. Few automobile mechanics were familiar with electric technology, and technical manuals, popular magazines, and training courses were non-existent. And, as with the earlier reluctant adoption of catalytic converters, electric vehicles were expensive.[46]

Electrics were also problematic as a solution to the problems of air pollution, health degradation, and climate change caused by internal combustion vehicles. Petroleum-fuelled designs had benefitted from a half-century of refinement. Yet careful attention was given to internal combustion engine efficiency only after the OPEC oil crisis drove up fuel prices (and declined again as heavy sports utility vehicles [SUVs] became fashionable). Conventional cars had long been conceived as technologies closely enmeshed with wider systems and social uses. By contrast, electric battery technology and operating controls, for example, were open to experimentation. Electric vehicles were also culturally ill-equipped to be the solution to an abstract problem of pollution. Early-adopter motorists after the oil crisis found them ungainly, unaesthetic, and impractical for lifestyles tailored to petroleum-fuelled cars. A typical but common frustration was how electric vehicle ownership could be made compatible with apartment dwelling when charging points were not available near parking spaces available to renters. Another was the need to accommodate charging times that were much longer than conventional fill-ups at service stations. Matters were worse for the wider urban community: even the relative silence of electric vehicles made them appear a threat to other road users habituated to larger and louder cars.[47]

A deeper and less often confronted problem is that electric car technology has not provided a rounded technical solution. The first generation of electrics at the turn of the twentieth century addressed the perceived

Figure 6.4
Electric cars for gender
equality. (a) Pope Waverley,
1907; (b) Columbia Model
68 Victoria Phaeton and
home charging unit, 1912.

problems of relative complexity, untidiness of operation, and physical labour needed to crank petrol engines. Those of the 1970s sought to bypass the newly identified problems of reliance on undependable, expensive, and air-polluting hydrocarbon fuel supplies. More recent electric car designs adopt a more ambitious but nebulous aim of being "greener" – that is, less harmful to wider environments. The term has been used inconsistently and has been appropriated by disparate interests. At one extreme, environmentalists may argue that "green" technologies should be sustainable in the long term: introducing no changes to the environment that cannot be compensated on the fly and relying on resources that can be renewed indefinitely and that respect existing cultures and human ideals. At the other extreme, corporations may conflate or redefine usage of terms such as "renewable" and "sustainable." By focusing on specific technologies, locales, or affected users rather than on the wider systems of which they are a part, such "greenwash" tends to downplay technical and social problems and to dilute potential solutions.[48]

In the case of electric vehicles, this downscaling of environmental analysis has typically focused on vehicle characteristics, while ignoring the environmental implications of its raw materials, manufacture, and energy supply. The State of California's Zero Emissions Vehicle Program, for instance, defined its objective as reducing exhaust gas emissions to zero. Yet relatively uncommon component materials for batteries and motors may be sourced from distant mines, and manufacturing processes may require piecework at widely separated factories. Extraction and refining processes and transport of components carries environmental implications. And while the vehicles themselves may not pollute, the source of their electricity usually does. Many, burning coal, peat, or even relatively cleaner natural gas, produce significant amounts of pollutants and generate carbon dioxide contributing to climate change. Thus, electric vehicles, conceived as standalone devices, have generally transferred the problem of pollution and climate change to large power plants. Components of innovative electric vehicles raise the same issues as more specific technological fixes such as catalytic converters: How can depleted batteries and broken rare-earth metal parts be sustainably reused or recycled? It is worth pointing out that, even for cleaner energy alternatives such as wind, wave, and solar power, the materials and manufacturing processes may not be fully sustainable:

in the early twenty-first century, few technologies are yet conceived as systems that are inescapably integrated with environmental and social systems. Electric vehicles, as conceived from the rebirth of research and development during the 1970s, have thus far been a partial technological fix that inadequately fosters sustainable modernity.[49]

Other problems of automobiles have attracted distinctive technological fixes. Alvin Weinberg, like Ralph Nader and Howard Scott and a generation of engineers, had suggested that both cars and roads could be designed so that safety is built in. They cited techno-fixes such as better formulations for tire rubber and treads to improve traction, stronger vehicle bodies to resist collisions, and road surfaces that reduce glare from headlights. Most strove to bypass human errors rather than to address means of reducing them directly.[50]

There were other equally unexpected and chronic consequences of cars at both the individual and collective levels. As discussed in chapter 3, the rationalization of city design had become a postwar focus for local governments. From the 1960s, urban planners increasingly identified new problems of city life caused specifically by the rise of car usage. Congestion of routes became acute in some cities, particularly in European centres that had inherited medieval town layouts. Language itself adapted to reflect the growing side effects of car travel. "Rush hour," coined with the arrival of the first motor vehicles at the turn of the century, quadrupled in usage between 1940 and 2000. The phrase "bumper-to-bumper" appeared before the Second World War to describe the slow crawl of a stream of cars along congested roads. And "gridlock" – a term entering usage as recently as the 1980s – emerged to describe how traffic flow could be jammed by cars blocking intersections in dense cities and even on highway systems.[51] As early as 1932, H.G. Wells had complained in a radio address of the lack of foresight in considering such side effects:

> The motor car ought to have been anticipated at the beginning of the century. It was bound to come. It was bound to be cheapened and made abundant. It was bound to change our roads, take passenger and goods traffic from the railways, alter the distribution of our population, congest our towns with traffic. It was bound to make it possible for a man to commit a robbery or murder in Devonshire overnight and

breakfast in London or Birmingham. Did we do anything to work out any of these consequences of the motor car before they came?

We did nothing to our roads until they were chocked; we did nothing to adjust our railroads to fit in with this new element in life until they were overtaken and bankrupt; we have still to bring our police up to date with the motor bandit.[52]

The rise of car ownership in cities such as Los Angeles revealed a restructuring and growing dependence of city life on automobiles. On-street parking narrowed roads further, and parking lots dedicated a growing amount of city real estate to motorists.[53] Growing reliance on cars caused a gradual attrition of public transport, a vicious circle in which urban workers became more heavily reliant on their private vehicles. Newer suburbs became increasingly dependent on car journeys as lack of sidewalk provision made some destinations inaccessible. Practical commuting to outlying districts required freeway travel that was prohibited to pedestrians and bicycles, and the freeways themselves were blocked by commuters not just at rush hours but periodically and unpredictably throughout the working day. The rising momentum of car usage by city-dwellers had locked in transport technology and encouraged alternatives to wither away.[54] In these constraining circumstances, the coincident rise of car culture was both a palliative for frustrated motorists and a pressure accentuating their problems.[55]

The postwar replanning of urban spaces had aimed to rationalize routes through cities but was limited by available funds and real estate ownership. Some quick fixes were a combination of legislative and technological responses. Parking meters, for instance, discouraged parking beyond the needs of shopping; parking laws prohibited street parking on the busiest roads; underground and multi-story carparks allowed denser packing of stationary cars. By the 1960s, however, cities were planning urgent large-scale technological fixes in the form of major new roads to circumvent congestion.

A commonly adopted technical solution was the creation of express routes and city bypasses, which reduced the bottlenecks of traffic signals and indirect routes. Such rational streamlining was understood as a technological fix, which accepted that the capacity of the road network would have to accommodate the uncontrollably rising social demand for car transport. But there was a flip side to this. City bypasses could force social change, too. They were often planned to pass through dense, low-income, and low-taxation districts and thereby require dislocation and rehousing of their residents. Thus, two social effects were forecast for a single technical solution: drivers would have faster journeys, and city dwellers would have better housing. Not uncommonly, however, some cities purposefully designed routes to eradicate declining business districts and slum neighbourhoods. Civic appropriation and compulsory purchase permitted the clearing of swathes of land for ground-level and elevated roads, translating the principle of intercity highways into dense city centres.

The postwar road infrastructure of Boston, for example, was planned as two circumferential expressways (Route 128 and Interstate 495), with a series of radial highways into the city itself. Implementation began in the late 1950s, and one of the radial freeways, a combination of elevated and underground sections known as the Central Artery, was completed.

In the United Kingdom, transport planners at the national level sought to match road capacity to anticipated economic growth, which was expected to correlate closely with car ownership. Over centuries of history, many population centres in the country had grown as destinations for markets and overnight travellers on longer journeys. Unplanned networks of roads between towns consequently captured traffic in circuitous routes and restricted flow. Planners sought to reduce delays with new bypass roads to circumvent through-traffic, as in the towns of Newbury and Dumfries. Larger, older cities such as Leeds were judged to require one or more ring roads to allow traffic circulation around, rather than directly through, their densely populated centres. Glasgow adopted a combination of shallow road cuttings, elevated fly-overs, and bypass roads to navigate the city. In much larger London, a combination of new roads, segregated access to pedestrian areas, and outright curtailment of traffic was deemed necessary.[56]

Yet these fixes typically failed to achieve their designers' aims and, indeed, alienated the putative beneficiaries. Complex sociotechnical systems adapt unpredictably to compensate step changes. New routes quickly became congested as drivers adopted them for quicker journeys or, even worse, encouraged more city-dwellers to become motorists, a phenomenon known as induced demand. Restricted access expressways were censured as contributing to urban decay in the areas they passed through or over. The loss of neighbourhoods, seen by some residents as more important than the newer replacement housing they were required to occupy, was increasingly resisted city by city. For more affluent audiences, such road projects were also opposed because of their prioritizing of traffic flow and commerce over the cultural value of historic districts, parks, and natural environments.[57] Public protests resisted the expansion of freeways in San Francisco as early as 1955 and, over the next fifteen years, the expressway projects in numerous cities in the United States, United Kingdom, Australia, and Canada were halted by public protests. In some cities, notably Washington (DC), New York City, Toronto, and Tokyo, plans were taken forward to dismantle the new roadways and replace them with more humane urban environments. In Boston, expressway development was stopped by citizens protesting the division of their neighbourhoods, traffic noise, reduced property values, and high motorist fatalities on the expressways. The state opted to build an underground road system (a project dubbed "the Big Dig") to improve public transport and to demolish the original solution. Even so, within a decade of the new road system's completion, traffic congestion matched its previous levels.[58] Taken country by country, the history of such fixes thus represents two kinds of failure: the technical inability to keep up with rising traffic and the legislative inability to encourage city residents to accept the necessary social adaptations. Collectively, the negative public reaction to expressways as technological solutions is the most dramatic rejection of technological fixes thus far seen in the modern world.[59]

Plastics and the Problems of Ubiquity

Still another range of technologies that came to represent twentieth-century modernity are plastics. One by one, plastic products formed into novel shapes and substituting for traditional materials were conceived as solutions for modern life, seemingly free from concern.

Bakelite, the earliest of these, was a sombre-coloured thermosetting substance patented in 1909 that proved ideal as a stable heat-resistant and electrically insulating material for the electrical industry. It became an intrinsic part of the experience of using telephones and lamp switches. Catalin, another casting resin introduced in 1928, provided brightly coloured consumer products such as radios, jewellery, and kitchen items. Celluloid, a product of cellulose from wood, cotton, or hemp, proved essential to the film industry and equally representative of the Machine Age. Cellophane, patented in 1912, became popular as a transparent airtight wrapping material during the 1930s. And patented by Du Pont in 1938, nylon was first displayed at the World of Tomorrow at the 1939 New York World's Fair and was sold commercially a year later for women's hosiery claimed to be "strong as steel." It was marketed initially as a synthetic textile based on the familiar components of "coal, air and water," a coy way of identifying a by-product of the growing petrochemical industry. Large chemical companies – Du Pont de Nemours, Monsanto, and Union Carbide and Carbon Corporation – integrated smaller firms during this period and plastics production accelerated. A bewildering variety of chemical products with competing claims – Celanese, Ductillite, Durez, Formica, Plaskon, Rayon, Styron, and Tenite among them – were available to middle-class consumers and other manufacturers by the end of the decade. The smooth, clean, and repeatable characteristics of these products epitomized optimistic modernity for interwar audiences (figure 6.5).[60]

With war looming for Americans, *Fortune* magazine forecast plastics as replacements for metals, wood, and natural fabrics for most domestic and industrial products. It predicted plastic gas masks, plastic gunstocks, plastic equivalents of tin cans, and even plastic aircraft. While these previews were optimistic, the Second World War made plastics familiar to members of the armed forces and more manufacturers. The scarcity of wartime materials encouraged substitutes. Polymethyl methacrylate (PMMA, marketed as Acrylic, Plexiglas, Perspex, or Lucite) became a tough replacement for glass in aircraft windows; nylon substituted for silk in parachutes and canvas in tents.[61]

Comparable shortages existed immediately after the war as firms struggled to satisfy pent-up demand for consumer products. Plastics exploded in popularity. Toys that had been made of cast iron, tin, or wood before the war were now available in plastic form. Packaging shifted increasingly from paper-based to plastic wrapping. Bathroom porcelain and metal plumbing,

Figure 6.5 *Opposite* Plastics and scientific modernity. (a) *The Plastic Age* (dir. Wesley Ruggles, 1925), based on a 1924 novel about modern campus life by Percy Marks. (b) Bakelite advertisement, 1938.

B.P. Schulberg

"The PLASTIC AGE"

FROM THE FAMOUS NOVEL OF YOUTH BY PERCY MARKS
DIRECTED BY WESLEY RUGGLES

CLARA BOW. DONALD KEITH

MARY ALDEN, HENRY B. WAITHALL, GILBERT ROLAND

PREFERRED
PICTURES

FROM PLASTICS HEADQUARTERS –

Over 2000 PLASTIC MATERIALS

JEWEL-LIKE solid materials of eye-intriguing beauty, or tenacious liquid cements for manufacturing processes...shimmering waterproof fabrics for costume design, or rugged materials that produce stronger abrasive wheels! As far apart as these plastics are, they merely suggest the sweeping range of more than 2000 materials available at Bakelite Plastics Headquarters.

There are lustrous Bakelite materials, in molded, laminated, and cast form, that lend sales-winning glamour to products. There are synthetic

resins that add endurance to paints and varnishes. There are cements, calendering and coating materials, wood adhesives, and bonding plastics for countless uses.

In basic composition, also, Bakelite plastics differ widely. They include ureas, polystyrene, acetate, and phenolic materials, in transparent,

translucent and opaque types, and in a color-range that rivals the rainbow. Their mechanical, electrical and chemical properties are equally varied, and meet a wide variety of technical requirements.

Save time, simplify your problem of material selection, by consulting Bakelite Plastics Headquarters first. Also write for Portfolio 27 of Bakelite booklets.

Bakelite Corporation, 247 Park Ave., New York, N.Y.
Bakelite Corporation of Canada, Ltd., 163 Dufferin Street, Toronto

BAKELITE

P L A S T I C S   H E A D Q U A R T E R S

Brittle-setting Cements for Brushes   *Cast Resinoid Cutlery Handles*   *Resinoid Bonds for Abrasive Wheels*   *Flexible Resinoids for Waterproof Fabrics*

kitchen surfaces, tableware and storage containers, lounge carpeting, chairs and light fixtures all found synthetic polymer equivalents and optimistic promotion. Plastic products could be lighter, cheaper, more durable, more hygienic, more aesthetically pleasing or lower cost than their predecessors. In short, the commercial rhetoric, echoed by consumers, claimed that plastics were the technological solution to the multiple problems facing wartime and postwar society. The problems with plastics have been slower to be appreciated than those of other technologies: the first criticisms were voiced a quarter-century after the war, and public concerns have become widely shared only during the 2010s.[62] Unanticipated side effects of plastics were discovered at multiple scales.

One of the first noted problems was the complement of the mechanical robustness of some plastics. Discarded plastic products may not rust, rot, or decompose like metal, food, and wood products. Aromatic polyesters, for instance, are little affected by weathering and microbes. As a result, plastic waste is persistent and increasingly obvious in waste sites and wider contexts. Gathered by wind and water, plastic cups and bags became ubiquitous along roadsides and watercourses by the 1960s. The visibility of plastic rubbish in urban environments was a trigger for action in early anti-littering and later anti-pollution campaigns. Nevertheless, the rising popularity of fast food and drinks exacerbated these problems through the end of the twentieth century.[63]

Among the most recently recognized aspect of this is the effect of plastics on marine life. Plastic reaches oceans through at least three routes: being washed out to sea from inland waterways or waste pipe discharges; unintentional spills of plastic intermediate materials from cargo vessels; and deliberate dumping of plastic refuse from ships. Immersed in cold seawater, such waste is even more long-lived than on land. A serious (and quite unforeseen) component of the problem is the mechanical assault of plastic materials on marine life. Discarded nylon fishing lines and nets, for example, estimated to take centuries to decompose, can become entangled on sea birds, fish, and marine mammals, impeding their movement and feeding and resulting in death. Plastic bags and food containers, among the most common waste products in seas, are similarly lethal. Microplastics, abraded and disintegrated by mechanical stresses and weathering, may be ingested by sea creatures and restrict nutritious feeding. Floating plastic particles resemble the zooplankton consumed by jellyfish; plastic bags resemble jellyfish for feeding sea turtles. As with wind and rainwater on land, sea currents concentrate this waste into large ocean gyres and deposit a portion on isolated beaches to further contaminate vast marine environments. Within these regions, particulates may exceed the normal food supply.[64]

A second problem not anticipated by earlier generations of enthusiastic adopters is that certain plastics *do* degrade in natural environments but in ways that make them the source of new problems. For example, Buckminster Fuller's promotion of geodesic domes to create capacious and strong buildings has fallen out of fashion, in part because the commonly used polycarbonate plastic panels deteriorate when exposed to ultraviolet-rich sunlight, causing leaks and loss of transparency. The rundown appearance of fogged plastic windows became familiar to counterculture commune builders and eager house-owners alike, who had adopted them from the 1970s. The degradation of the plastic caused by burning or exposure to water at room temperature also causes leaching of toxic bisphenol A (BPA), which, at high levels, can disrupt the endocrine system. This unexpected characteristic has been unusually divisive for technological enthusiasts. Dome-builders have retrenched to promote more complex building solutions; investigations by researchers funded by the plastics industry have tended to downplay concerns, and government-funded investigations highlight them. BPA and polystyrene in sea-borne plastics can also leach out and concentrate on waste accumulating at the sea surface and on beaches, endangering marine life. The accumulation of such chemical constituents in individual species and propagating throughout the food chain is a more general case of the problems first discussed by Rachel Carson.[65]

Intentionally biodegradable plastics, identified as a potential technological fix for plastic litter during the 1970s, also carry caveats. So-called "compostable" or "environmentally degradable" plastics decompose into organic molecules such as carbon dioxide and water; "biodegradable" plastic material similarly breaks down until it is visually indistinguishable, but it also breaks down into inorganic subcomponents. Compostables typically made from biomass such as corn starch are a distinct minority: there are fewer options for such synthetic materials having the desired manufacturing properties. More abundant plastics are synthesized from hydrocarbons originating from petrochemicals. Biodegradable chlorinated plastics such as polyvinyl chloride (pvc), for instance, decompose to release toxic chemicals that can reach dangerous levels when concentrated in groundwater near landfill sites. The fortuitous discovery and use of bacteria that can consume plastics is also of mixed benefit. Flavobacteria consume polymers such as nylon, for instance, but produce methane as a by-product: a greenhouse gas much more damaging than carbon dioxide. While other microbes have been identified that generate plastics from methane, such bioengineering – and its associated complexities and side effects – are comparable to those of the plastic starting material.[66] A recently promoted variety of non-compostable material is oxo biodegradable plastic, in which oxidation breaks down the material into organic chemicals and hydrocarbons, some of which can then be consumed by bacteria. These degrade faster than more common plastics but endure months or years in their original form, and longer as fragmented plastic pieces. The European Commission has concluded that oxo-plastics are likely to biodegrade more slowly than claimed in natural environments and to generate microplastics that endanger marine life. As with other technologies of commercial importance having identified side effects, manufacturers have aggressively opposed the conclusions of government-funded researchers and attempts to restrict usage of the products.[67]

An alternative technological approach has been to devise plastics that can exploit intentional photo-degradation. For example, polypropylene, produced commercially from 1957 and now one of the most widely used commodity plastics, and Kevlar, a synthetic fibre developed in 1965, were recognized soon after their introduction to break down when exposed to the ultraviolet (uv) component of sunlight. This characteristic makes them unsuitable for products used outdoors unless enhanced by uv stabilizers (these additional chemical components, like others introduced into plastics formulations to alter their physical properties for specific applications, carry their own particular environmental implications).[68] Purposeful uv degradation can be targeted as a fix for plastic waste in the environment. However, such exposure tends to reduce the mechanical strength of the materials without eradicating them completely. They may also decompose into inorganic constituents. Incineration of plastics, unless at very high temperatures, also releases compounds such as dioxins into the air. Decomposition – either via biodegradation, photo-degradation, or incineration – can, at best, redistribute and dilute the chemical constituents of plastics based on petrochemicals. None of these technological options provides a straightforward fix to allow the continued proliferation of plastics in modern society and preservation of current lifestyles.

Just as plastics have evolved as a jumble of different chemical products, means of recycling them also pose considerable technical problems. Recycling is itself a technological fix that reveals necessary, but often overlooked, social accommodations when investigated further. Recycling requires separating different materials so that distinct chemical processes can be applied to each. Milk jugs (high-density polyethylene, or HDPE), soft-drink bottles (polyethylene terephthalate, or PET), and cooking oil bottles (PVC) are reusable, recyclable, and non-recyclable, respectively. Polystyrene (PS), one of the most common materials for Styrofoam drinking cups, take-out foods, and plastic cutlery, cannot be recycled economically and makes up a significant portion of landfill material in the United States.[69]

Arguably, the cultural confidence in plastics, and expectation of a sustainable solution to the problems they create, delayed other responses. Social and legal measures for controlling such waste have multiplied since the 1960s. These include anti-littering legislation aimed at consumers, clean-up campaigns organized by community groups, and, more tardily, laws mandating social responsibility on the part of manufacturers and distributors. These have discouraged use of certain plastic materials and products such as plastic bags by imposing a mandatory charge on purchasers. Legislation has also encouraged segregation of waste and recycling, a responsibility shared between householders, local government services, and plastics companies. Interestingly, the more fundamental approach to the problem – minimizing plastics manufacturing and waste in the first place – has been among the latest and least pursued of options. Like DDT and cigarette filters, the problems with plastics have been highlighted by citizens, governments, consumers, and manufacturers in that order (while consumers and manufacturers are, of course, citizens, interests are commonly compartmentalized, placing employment and profits higher in the hierarchy of attentions).[70] The solving of the problems created by plastics, more often than not, has been addressed by changing cultural norms about "fashionable consumption" rather than by adopting technological solutions.

There are consequently two negative outcomes traded off in the usage of common plastics: persistent plastic waste that threatens lifeforms via what could be called "bio-mechanical" effects, on the one hand, and decomposing plastic waste that releases chemical constituents that are environmentally toxic or contribute to climate change, on the other. The technological fixes for the problems of plastic do not yet adequately address the fundamental problem of what to do with the starting materials.[71]

Beyond these physical effects, there have been wider societal implications. The proliferation of plastics has been aided both by complementary technologies and cultural choices. Consumer adoption of microwave ovens from the late 1960s, for example, began a trend by manufacturers to develop convenience foods packaged in plastics compatible with microwave heating. Purchasers progressively selected convenience over other concerns. Palatability and choice of microwavable foods, which were initially poor, thus gradually improved a generation later. Busy homes, now striving towards all adults in paid work, gratefully dispensed with the demands of cooking from scratch and stocking a wide range of ingredients. Similar social and cultural changes encouraged the growth of fast-food dining, which placed growing reliance on non-recyclable Styrofoam containers and plastic-reinforced cups. More recently still, the fashion for mobile drinking – once culturally discouraged – caused an explosion in disposable water bottles, coffee cups, and covers. These societal trends in food consumption, only a small part of the cultural shifts of the late twentieth century, dramatically escalated the quantity and variety of plastic waste.

Other social choices exacerbate the technological problems. Plastic products typically cannot be repaired, as can conventional wood and metal components. Broken or outmoded items are thus more commonly discarded than re-purposed. In turn, throw-away culture accelerates consumer demand for new and novel products, while simultaneously magnifying the side effects by generating more toxic and durable waste. Collecting glass bottles for bottle banks, for instance, was a source of pocket money for previous generations of children, but this has been eclipsed by growing mountains of single-use plastic bottles. Similarly, consumers' perceptions of the lack of reparability of their purchases discourages the flourishing of entire classes of artisanal expertise, ranging from cobblers' mending of shoes to seamstresses' clothing adjustment to lawnmower and television repair shops. These have domino-like effects in collapsing networks of expertise, formal training programs, manual trades, and complex local economies. The collective choice of plastic as a ubiquitous material consequently reinforces cultural belief in progress while streamlining available alternatives.[72]

Even so, societal changes have challenged this seemingly inevitable effect of new technology. The trend in mass consumption facilitated by the adoption of plastics has had distinct cultural reactions in the early twenty-first century. One is the rise of "retro" culture, in which a younger generation opts for older products and styling, and their older counterparts choose to retain some of their familiar technologies. Examples include consuming vinyl records instead of CDs, mp3 downloads and streaming music, and adopting clothing and fashions of the past. This fashion trend echoes the reaction to factory-produced goods in the late nineteenth century that led to the Arts and Crafts movement supported by William Morris and John Ruskin in Britain. Their ideals resisted the uniformity and ubiquity of machine-made products in favour of hand-made goods and traditional craftsmanship. This return to premodern methods of fabrication hints at a deeper cultural investment in the social aspects of technology. Nevertheless, the implications of retro fashion are not clear-cut. Some products of previous generations – notably vintage automobiles and diners – were showpieces for technologies now identified as the source of serious side effects. Valorizing products of the past can literally reproduce the economics and consumption patterns that drove their original adoption. In the same vein, manufacturers have rapidly adapted by marketing products that simulate the aesthetics of older technologies. The attraction may also be skin deep: most consumers do not associate repair and repurposing with the retro products they choose.[73]

An equally problematic social adaptation is the self-production of products. Under the rubric of the "maker movement," proponents reject some mass-produced consumer goods to make items that serve their personal needs or express individual creativity. While some of this relies on traditional skills such as knitting and woodworking, much is reliant on new technologies, too. Computer-control and coding are key components for some makers as these allow personalization and seemingly unlimited creativity. On the other hand, there has been little discrimination among enthusiasts regarding the environmental implications or, indeed, the general side effects of their activities. 3D printing, for instance, relies on polymers to generate software-defined objects and may rely on industry-supplied materials;

adopted on a large scale, this threatens to reproduce the historic problems of plastic production and waste, and to reinforce its corporate direction. A related ambiguity is that the purposes for engaging in these activities are just as varied among makers as in they are in the wider society. Some participants identify an antipathy to certain technologies and/or their social implications, but others are just as eager to explore and invent as were their early twentieth-century counterparts in engineering firms and science fiction fandom. As a result, it is unclear whether the maker movement offers an alternative to the technological side effects of modern society or merely reproduces them on a finer-grained scale that may evolve over time. These social trends, initially appearing to counteract the excesses of modern technologies, can thus be appropriated and subverted.[74]

Common Features of Unanticipated Problems

From their 1973 studies of methadone, Antabuse, instructional television, breathalyzers, and intrauterine devices, Etzioni and Remp arrived at two partial insights. The first was that the applicability and success of technological shortcuts could seldom be assessed cleanly. The initial conditions and contributory factors of problems were usually complex, and proposed solutions tended to work unevenly across different social contexts. Their second insight was that social solutions proposed as alternatives faced similar problems of analysis.[75]

A complementary approach is epitomized by Edward Tenner's popular book *Why Things Bite Back*, which narrates tales of the frustrations of designing technologies to solve our problems. He coins the term "revenge effect" to describe how technological interventions can be negated by unanticipated counter-effects. The case of cigarette filters – which encouraged some to take up or to continue smoking – is such an effect. A similarly recognizable phenomenon is what he terms "recongesting effects," which follow innovations designed to increase capacity. As discussed above, the rapid increase in car usage that followed the introduction of motorways and town bypasses in the United States and Britain from the 1960s made these urban improvements only temporarily beneficial and exacerbated the need for still more capacity. Tenner suggests that modern technologies tend to be good at solving acute problems but, in turn, generate new chronic problems. Such descriptive categories are salutary but offer few insights into better sociotechnical approaches or into the qualities of sociotechnical systems themselves. He provides good examples of *how*, but not quite *why*, technologies "bite back."[76]

Given the difficulty of forecasting and of fully understanding the consequences of problem-solving, analysts are restricted to empirical studies of past cases. This approach has its faults. Points of similarity may be difficult to discern. Developing a typology of social, political, and cultural consequences may appear hopelessly more complex than even categorizing typical forms of engineering error. Such attempts at analysis and forecasting can also be readily countered by the claim that empirical cases have provided insights to problem-solvers that will prevent similar mistakes in the future. The most effective rebuttal to such assurances is the citing of past and present cases of failed fixes to demonstrate their steady, if not growing, frequency in modern societies.

The historical cases outlined in this chapter evince some shared qualities and recurring themes, which are developed more abstractly in chapter 7. First, and to varying degrees, they were driven by optimism: intense, uncritical, and (in retrospect) unjustified optimism. They expressed cultural confidence in the powers of science and engineering, and in the trustworthiness of authorities. The problem of excessive faith in technology was paralleled by trust in elite experts and paternalistic government. If DDT had proven effective in saving soldiers' lives and winning the war, what better commendation could there be for employing the same wonder pesticide on farms and in homes? To a lesser degree, and more surprisingly, in retrospect, cultural confidence in innovations by companies was equally high. Cynical cigarette firms could tout the improvement in "throat feel" of filtered cigarettes and find that their customers uncritically accepted that filters would remove all the harmful components later discovered to be the source of chronic disease. For wholly new products associated with the modern age, consumer faith has proven almost impossible to temper. Our engagement with plastics – initially via the appealing Bakelite texture of telephone receivers, later in the first generation of children's toys, and more recently in microwavable containers – has generally been eager and indiscriminate. Plastics usage continues to rise for home conveniences, even as waste dumps plough most of it, "recyclable" or not, into land fill.

A second shared quality is that these cases are examples of short-sightedness on the part of designers, legislators, and consumers. They typically identified a significant benefit in the short term: pest-free crops, safe smoking, convenient journeys, or durable synthetic materials – in short, material advantages at affordable cost. Engineers responsible for these innovations often genuinely identified immediate gains without pausing to examine longer-term outcomes. This is a crude form of design relying on first-order analysis, which seeks to achieve a desired goal by successive approximations. Legislators, encouraged by innovating firms that entrained rising employment, growing economies, a better standard of living, and largely contented consumers, identified positives for all. The unanticipated effects for other audiences (including non-users or other species) were seldom considered.

Third, these cases exhibit chronic misdirection of attention. Inexperience with regard to novel technologies and their potential side effects exacerbated such complacency. Problems, when they became apparent, developed years after introduction and, typically, were discovered by experts in other fields. In a disturbing number of cases, manufacturers and distributors of products sought to downplay side effects to preserve their markets. This deliberate skewing of the record is not a necessary outcome of technological solutions, but it tends to prop up public confidence in them. The importance of larger firms and corporations for national economies has too often encouraged government collusion in the form of tax dispensations, direct funding, and relaxation of regulation. And working communities themselves have frequently opted for short-term needs such as jobs instead of long-term gains such as sustainable environments for future generations. Such cases of demonstrated harms from technologies are not so much a celebration of endless progress as a Faustian bargain to make the most of the present.[77]

I argue that these themes are generic and are not limited to a single country's constellation of companies and particular government legislation. They resonate with modern cultural attitudes now widely shared in many countries. The historical cases have been sketched to provoke readers' sensitivities to persistent lapses of critical ability. None of them represents merely naïve consumers of a simpler age; nor can they be read simplistically as tales of an under-legislated era.

The most important shared attribute of such cases is the difficulty of foreseeing how they would prove to fall short. Each technology – DDT, cigarettes, plastics, and car engines – exploded in popularity decades before side effects were tentatively reported and acted upon. Anthropogenic climate change, a life-threatening side effect having global proportions, was belated confirmed more than two centuries after large-scale industrialization began. Most technological shortcomings, in fact, have not yet been corrected, even if notions of predictive uncertainty and mixed outcomes of innovation have long been appreciated. A "law of unintended consequences" was first discussed in relation to economics: John Locke (1632–1704) specifically considered the uncertainty of the effects of regulating interest rates, but the sociologist Robert Merton (1910–2003) provided the first modern summary in 1936, showing that human expectations and behaviours can lead to paradoxical outcomes. However, the historical exploration of technological fixes extends this caution to the still-unfamiliar social territory of technological choices. Technological solutions *routinely* have unintended outcomes, and their temporary utility may be swamped by belated negative side effects. Despite our shared cultural faith, historical cases show that technological innovation is not reliably positive in outcome and tends to be supplemented tardily by familiar social, political, and cultural measures.[78]

7 Implications of Technological Confidence

Reassessing Technological Hopes and Dreams

The first five chapters of this book chronicle the rise of technological faith among distinct social interests, and the sixth explores some important cases of their later-developing side effects. This historical examination can be taken further. I have explored this sociology of expectations and contrasted it with wider-ranging consequences reported by medical researchers, environmental scientists, and citizen-activists. The marshalled evidence suggests that technological over-confidence is a recent cultural turn having largely negative outcomes. This contrasts with the dominant narrative from business and often government, which argues that "future expectations and promises are crucial to providing the dynamism and momentum upon which so many ventures in science and technology depend," even while acknowledging that failed expectations have "severely damaged the reputation and credibility of professions, institutions and industry." Such unbalanced attention from techno-fixers does not address the general case of characteristic outcomes from typical technological problem-solving.[1]

The present chapter more overtly flags generic features of these and other historical contexts, and common consequences of over-confidence in technological fixes. It identifies recurring themes, assumptions, and rhetoric shaping modern engagements with technologies.

Instilling Unrealistic Optimism

From the Technocrats to Alvin Weinberg and recent Silicon Valley entrepreneurs, enthusiasts have argued that technologies provide quicker fixes
to human problems than do conventional social, political, and economic
options. An important corollary to this is the assumption that there is a
straightforward cause-and-effect relationship at play: that a technological
intervention will deliver a deterministic solution while leaving other variables unchanged.

In a fast-paced modern world, goes the argument, trust in technological
solutions is the rational choice. This "can-do" engineering perspective is un-
derpinned by the further conviction that approximate fixes, however crude,
can be cleverly refined by subsequent fixes. If a technology has side effects,
this suggests, it can be corrected by timely modifications as the faults are
gradually discovered. Iteration should yield incremental improvements that
converge towards a better product.

The history of software engineering underlines how this design philosophy was mainstreamed during the late twentieth century. Throughout the
Second World War, designers typically engineered commercial products to
be viable offerings that, once released, either sold successfully or were abandoned. Mass production manufacturing processes necessitated stable parts
lists, assembly jigs, and worker tasks. In the burgeoning postwar period of
commercial innovation, however, production-line organization adapted to
consumers, who came to expect that design faults in products would be corrected by replacements from the manufacturer. In important respects, the
software boom that followed the introduction of mass-market personal
computers in the 1980s altered these conventions, introducing a higher cultural tolerance for technological faults. Consumer software products were
considerably less familiar to designers than were older technologies, who
consequently delivered their products with more undiagnosed faults. Indeed, the computer industry had rapidly expanded without the traditions
of engineering practice or standardized training. The term "software engineering" was introduced only in 1968; the earlier terms for the nascent profession, common in mainframe computer companies such as IBM and GE,

were "programmer" or "systems analyst." Software design had emerged from mathematics, as in its wartime development by Alan Turing, who had conceived it as a table or list of consecutive logical instructions. The term "software" itself, barely used before the 1960s, had begun as an in-joke used by electrical engineers to contrast it with the more serious engineering science of "hardware" design; the later term "firmware," less familiar to consumers, similarly began as an in-joke a decade later to signal an uncomfortably in-between product that embedded software in permanent silicon devices.[2]

With these complex products came unpredictable side effects. As a young engineering art, software design lacked rigour and ready diagnosis. From its origins after the Second World War, the history of the computing industry was characterized by unrealistically optimistic forecasts and unanticipated failures of large-scale implementations, bringing even major companies close to commercial collapse.[3] The Year 2000 (Y2K), or Millennium Bug, was one such panic, fortunately incorrect, that the industry-wide digital memory shortcuts for date recording would cause unpredictable errors in commercial accounting and scheduling systems at the turn of the twenty-first century, leading to catastrophic failures.[4] Similar concerns about software products encourage permanent vigilance to rapidly repair the intentional exploitation by hackers of design loopholes that compromise security. The endemic over-confidence and technical failures in the industry were castigated by Fred Brooks (1931–), who had overseen the development of the IBM System/360 computer system during the 1960s. Brooks argued that promises of progress on the part of the computer and software industry were invariably overstated and that the human systems required to develop projects were always underestimated. His insider's account reiterates insights drawn from the examples in chapter 6, which reveal the economic, political, and cultural actions needed to support seemingly straightforward technological solutions. Brooks referred to this false promise as a "silver bullet" akin to the mythical solution to how to kill werewolves. Such a magical solution to a complex problem had close affinities to the more down-to-earth notion of the technological fix. The supernatural connotations of silver bullets, however, better suggest the unrealistic expectations of such modern folk beliefs about fixes.[5]

Software culture mutated notions of progress, too. The claimed progress of software technology became linked with new patterns of development.

Tools for monitoring the actions of programs were initially rudimentary and often devised as operational fixes by the programmers themselves. Consequently, commercial software producers adopted the convention of "bug fixes" from the late 1970s, replacing early software versions with updates either regularly or as emergency corrections for functional weaknesses. In the process, this survival tactic of inexperienced or under-resourced firms was converted into a widely shared cultural value: consumers came to appreciate claimed improvements to the usability of their imperfect microprocessor-controlled products as virtues and to identify this iterative correction as genuine progress. The now commonplace term "version number," incremented with encouraging regularity, redefined fault correction as evidence of inexorable improvement and manufacturer commitment. The link between software and progress thus became a truism.[6]

Short-Term Solutions

Confidence in problem-solving, whether technological, social, or political, can engender myopic solutions. The rapidly evolving software industry illustrates this generic feature of technological fixes: practitioners attend to immediate and specific problems rather than to larger system-level faults. Bug-fixing is driven by bug reports, and it deals with identified faults as they are discovered rather than seeking to analyze deeper design flaws and vulnerabilities. In a context of rapid product cycles and perpetually inadequate resources, this shallow attention prioritizes quick patches and forces an abandonment of robust design in favour of perpetual triage. Indeed, the term "software patch," adopted by the software industry in the mid-1970s, echoes Alvin Weinberg's description of "band-aid solutions."[7] Coupled to expectations of progress, this hasty cycle of development and support narrows the attention of those responsible for new technologies.[8]

This time-limited attention carries broad implications. On the technical level, fixes are unlikely to address systemic issues, or what computer architect Fred Brooks called "essential complexity." He argued that engineering tends to focus on "accidental complexity": immediate crises that are a result of short-term planning or shallow analysis. Ambitious designs, he suggested, require cautious analysis in advance or face the likelihood of being hobbled by subsequent fixes and patches. Countering expectations, this method of problem-solving proves more expensive and time-consuming, and more likely to spawn further unexpected outcomes. This culture of development and support may become self-perpetuating, restricting engineers' ability to see beyond the design horizon before the next unanticipated side effect comes along. Unrestrained technological enthusiasms thus engender failures, not successes.[9]

Technological fixes may appear particularly attractive to government policy-makers owing to their speed of implementation, their seeming simplicity, and, not least, their appeal to voters. Longer-term solutions are more problematic as election cycles may replace administrations before projects can be thoroughly planned and executed. In this myopic decision-making context, the complexity of sociotechnical systems is likely to be underestimated. Larger systemic change may be deferred or avoided altogether.

A prominent example of this point is government inaction with regard to organizing long-term solutions to radioactive waste. Cold War production of nuclear weapons took place at high-security installations before safety regulations were developed and, indeed, generally continued to bypass civilian standards. Radioactive by-products of plutonium extraction processes, for example, were not uncommonly vented to the atmosphere through chimneys; emissions were timed by weather forecasts of prevailing winds and personnel were advised to remain indoors. Miscellaneous liquid wastes have been stored in onsite tanks, leading to an unidentifiable and unstable soup of toxic ingredients combining in unpredictable ways, leaking to groundwater after decades of storage, and resisting further chemical neutralization. Civilian nuclear reactors created better-understood waste products but have had no generally established mechanism for ultimate disposal.[10]

Like software, nuclear waste has attracted numerous technological fixes seeking to reduce human and environmental dangers.[11] One class of fixes involves nuclear processes to convert radioactive substances into less toxic materials. Artificial transmutation irradiates materials in a particle accelerator to convert radioactive components into less radioactive products. An alternative is chemical reprocessing to extract plutonium and/or uranium from spent fuel, or nuclear reprocessing to recycle the recovered materials in breeder reactors to generate further energy. Relatively few sites reprocess fuel in this way, requiring its transport by rail or truck within countries or by sea between continents, which has increased public concerns about potential accidents. Breeder reactors as experimental closed systems were among the earliest hopes to generate practically inexhaustible energy, but they were plagued by operational problems that made them complex and uneconomic.[12]

The more popular fixes so far discussed concern storage. Spent uranium fuel elements, for instance, are typically stored in large pools of water in guarded facilities and sometimes onsite. Such storage near fragile environments is problematic. One of the concerns in the 2010 Fukushima accident was that a spent-fuel pool was *on top* of one of the reactor buildings that caught fire, cascading potential problems of reactor overheating and breach, radiation-bearing smoke, scattering of highly radioactive fuel rod materials in the wreckage, and leakage of irradiated cooling water into the sea.[13]

More permanent storage fixes, few trialled or adopted on a large scale, are more innovative. Some are controlled versions of the former military waste tanks. After a year or more spent cooling in pools, fuel rods may be enclosed in welded above-ground steel containers bathed in inert gas. Alternatively, fuel elements may be converted into more physically resistant materials. For example, their components may be "vitrified," or converted into a chemically resistant and water insoluble glass or ceramic, or rendered into a concrete-like material. More ambitiously, spent fuels may be deposited in inaccessible locations where they are expected to be free of human interference and environmental harms. Proposals have considered ocean floor disposal in the deep sea, deposition in disused salt mines deep within mountains, the use of deep boreholes or natural fissures likely to be geologically stable, and even space launch to send nuclear waste towards the sun. Each presents identifiable risks – such as need for periodic monitoring and maintenance, potential contamination of groundwater, vulnerability to earth tremors, or failure of launch vehicles – that have convinced governments to delay decisions.[14]

A physically and logistically larger problem has been the decommissioning of reactors themselves. Early plans for military reactors had been to bury or prevent access to the isolated facilities without attempting dismantlement. Later civilian planning envisaged collecting and segregating dangerous materials in safe locations. An example is the full decommissioning of Scotland's Chapelcross nuclear power station, one of the first internationally inaugurated (in 1959) and operating for forty-five years. It is estimated that it will take until 2059 – a full century after start-up – to return the site to other uses. The process includes removal of fuel elements, dismantling of radioactive parts, and disposal of other dangerous materials (such as some thirty-three hundred tons of asbestos used to insulate the hot reactor buildings and heat exchangers). In the interim, British facilities are guarded by a dedicated police force, the Civil Nuclear Constabulary (CNC). Details of both the decommissioning process and operation of the CNC, like those of the world nuclear industry as a whole, are matters largely beyond public scrutiny. Thus, a convenient technological fix for the problem of derelict nuclear power stations does not exist; the solution requires elaborate political planning, management of citizen expectations, and operational commitments over many decades.[15] As noted in chapter 3, Alvin Weinberg, the champion of both nuclear power and technological fixes, was troubled by the emerging realities of nuclear power. While he admitted naïvely forecasting nuclear energy "as a magical panacea" and "symbol of a new technologically oriented civilization, the ultimate 'technological fix' that would forever eliminate quarrels over scarce raw materials," he later decided that social design would also be essential. Safe exploitation demanded large reactor facilities and waste disposal sites far from population centres, to be tended by a vigilant "nuclear priesthood" that would instil public respect for the mysterious technology long after the power plants and other social institutions could be expected to survive. As he acknowledged, the societal demands were, in fact, unparalleled in human history.[16]

It is salutary to compare such protracted and complex outcomes to the initial promise of nuclear energy. Its development in Britain was undertaken and directed for a quarter-century by a government body, the UK Atomic Energy Authority (UKAEA), and subsequently managed by private companies. British nuclear power, like the experience in other countries, was in retrospect a hastily adopted technological fix for postwar political and economic problems. Successive UK governments hoped that the new source of energy would skirt the issue of national coal supply being periodically limited by striking miners and would foster a new export industry in a country nearly bankrupted by the Second World War. The first generation of reactors, such as Chapelcross, were also designed and operated to provide plutonium for nuclear weapons, a product deemed equally important by the

British government to assure the country of high status in international politics. Repeated with international variants over a half-century, the history of nuclear power is the story of an exciting technological fix gone bad.[17]

Selective Attentions

While the insights of senior managers like Fred Brooks profited from their overview of entire industries, they were still shaped by the dominant culture of technological fixing. Long-term considerations beyond their organizations were rare. Managing product design goals was their recognized remit, but responsibility for conceiving future labour supply and skills training was seldom considered. More important, the benefits and potential harms of their activities were narrowly conceived.

Shallow attentions have not been restricted to the software industry, nor are they a product of technology businesses alone. As evidenced in previous chapters, they are a widespread feature of modern technology-confident culture. The economic bottom line of technology firms does, though, clarify processes that may be more difficult to discern in government decision making and citizen choices. Consider the attention of businesses to their customers, for example. A company is likely to prioritize particular interests, which are usually short term and local. Most resources will be dedicated to current and prospective customers, and somewhat fewer to past customers or potential customers in the more distant future, say a decade hence. With business viability an ever-present threat, firms necessarily prioritize near-term customers over more distant ones. More importantly, they are likely to pay less heed to other interests, such as non-customers or non-human species potentially made vulnerable by their products. Commercial survival may depend on maintaining an existing customer base and income stream and so work against solving unexpected problems. In the worst cases, such as the companies discussed in chapter 6, this attention to corporate goals and near-term outcomes may also encourage their challenging of reported harmful side effects of their products via consumer marketing, the courts, or government lobbying.

In more muted terms, governments have followed similar strategies. Typically elected for a term of a few years, they require short-term responses to immediate problems that are of concern to their electors. Social sectors

without a direct influence on government problem-solving may consequently be disadvantaged. Traditionally underrepresented interests include
the poor (having little economic power to buy and influence products) and
the voiceless (notably non-voters, unborn future generations, and non-
humans). A widely repeated example throughout the twentieth century
was the funding of hydroelectric power by local, regional, and, eventually,
national governments. In fact, hydroelectric facilities became the most popular public utilities owing to earlier concerns about the operating practices
and excessive profits of private power companies. The earliest American
examples were built in the 1880s mainly to power electric light bulbs and
motors, and were rapidly extended to supply electricity to towns located
near mountains with fast-flowing rivers and valleys that could be flooded
to serve as reservoirs. Benefits of electric power were obvious; electric trams
and cars, the most common type at the turn of the twentieth century, relied
on such dependable power sources, and a burgeoning market for home appliances included newly introduced conveniences such as electric irons
(1903), toasters (1909), and refrigerators (1913) for well-off consumers. By
the First World War these were supplemented by luxuries such as curling
irons, corn poppers, and electric massagers.[18]

The earliest opposition to large dams came from those same privileged
audiences that had not only electricity but also extended education and a
political voice. For instance, the original intent for the Hetch Hetchy Dam
in California was the creation of a reliable replacement for San Francisco's
water supply following the city's devastating fire in 1906, but it was also
designed to supply hydroelectricity. The scheme was opposed by John Muir
and the Sierra Club, the major players in the creation of the American
National Park system, on the grounds that untainted reserves were needed
to protect nature from the inexorable expansion of modern cities and their
corruptions. While these spiritual, aesthetic, and social sensitivities were
not identical to modern environmental concerns, opponents of the dam
provided an unusually early example of technological oppositions for non-
technical reasons.

Hetch Hetchy identified themes that were later echoed in opposition to
nuclear plant cooling towers and wind turbines. By offending cultural
notions of aesthetics and the sanctity of pristine nature, technological
solutions such as power dams were claimed to cause unquantifiable but

significant harms. Although little acknowledged by decision makers at the time, the creation of the Hetch Hetchy Dam, and the surrounding Yosemite National Park founded in 1864, contributed to the removal of Native American hunter-gatherers who had for centuries made the valley their home. This eviction was a part of the large-scale intentional and involuntary redistribution of peoples and power on the continent. The technological imposition of the dam was thus built upon earlier political fixes that included the Indian Wars and treaties, the establishment of reservations, and a policy of exterminating American bison. The federal government judged in 1913 that provision of water and electricity to San Francisco outweighed the harms caused to the natural environment and marginal communities. Recent arguments for retaining the dam include the economic and environmental costs to dismantle it, and its value in discouraging visitors to overcrowded Yosemite National Park. The history of the dam and its controversies thus illustrates how narrow perspectives define and resolve human problems via technology.[19]

Governments came to argue that, despite its being limited initially to select audiences (like most newly introduced technologies), hydroelectric power was crucial for local and national economies. Firms, from electric car manufacturers to later vacuum cleaner and radio makers, flourished as new business opportunities were identified. Later, larger-scale projects identified further technological fixes. Dam projects could, of course, provide electricity for nearby communities, but high-voltage, long-distance transmission lines could distribute it across entire regions. The availability of electricity in rural areas opened them to new businesses, expanding the economy. And with the advent of radio broadcasting after the First World War, governments were quick to appreciate its value for cheap and immediate public education, particularly for isolated farmers and rural communities. By the 1930s, most industrialized countries were undertaking projects to electrify their rural areas into national networks. A second class of benefits for rural modernity was the ability of dams to regulate water supplies, distributing scarce water for agricultural irrigation and controlling flood surges in times of plenty. And a third class, again crucial to agriculture, was the industrial production of fertilizer via electrically supplied factories. Thus, large national projects such as the Tennessee Valley Authority (TVA), begun at the peak of the Great Depression by the incoming Roosevelt administration in 1933, sought to

Figure 7.1. Technological fixes for social sustainability. Tennessee Valley Authority
Pickwick Dam under construction, 1936.

combine technological fixes having multiple societal benefits (figure 7.1).
Most opposition came not from consumers of these benefits but from companies trying to compete with the lower-cost government products and
services. As both sides acknowledged, such technological fixes could skirt
conventional economics and transform socio-economic options, but the
outcomes were judged differently by different interests.[20]

Where the Hetch Hetchy disagreement had been atypical for its times, a
much larger opposition, setting the tone for subsequent projects, faced the
construction of the James Bay Hydroelectric Project a half-century later. A
vast hydroelectric system of dams, power transmission lines, and river diversions in northern Quebec, Canada, the project raised controversy from
its planning stages in the early 1970s. It transformed environments inhabited
by Cree and Inuit peoples and profoundly altered their means of gaining
livelihoods, engaging with their environments, and pursuing traditional
ways of life. Aggravating their justifiable grievances was the implementation
of the hydroelectric system principally to supply markets in southern
Canada and for onward sale to American markets. There were thus multiple
scales of inequity and injustice embodied in the design: inadequate consideration not only of Native Canadians but also of the national needs of Canadians versus Americans as well as site-by-site decisions to sacrifice the
well-being of local species and the larger ecosystem. This technological fix,

argued its opponents, in fact caused much more serious and irreparable harms than did the problem of inadequate electricity supply.[21]

From a handful of critical studies during the 1960s, the wider sociotechnical contexts of hydroelectric dam projects have increasingly been questioned by environmental scientists, scholars, and citizens' groups. New dam projects are now recognized as politically contentious: restricting water to neighbouring countries or regions, such as with the Grand Ethiopian Renaissance Dam across the Nile; environmentally problematic, such as the mammoth Three Gorges Dam across the Yangtse River in China; or socially disruptive, such as the Belo Monte Dam in Brazil blamed for forcing the displacement of thousands of Indigenous people. These concerns have shifted the balance of decision making from unalloyed optimism about such fixes to caution and contestation.[22]

So, like expressways through cities, hydroelectric projects underline the broad drawback of narrow attentions. They tend to be flawed examples of utilitarian thinking, targeting the "greatest good for the greatest number" inappropriately. The "greatest good" might be portrayed, seemingly unproblematically by the majority of citizens or power-holders, as faster roads or ample electricity, while overlooking lesser problems of uprooted communities or blockage of fish migrations. The "greatest number" is also likely to be conceived as regional or national populations instead of minority social groups and local circumstances. This trading-off of alternatives is applied unfairly to distinct parties in distinct circumstances. Thus, urban dwellers' provision of electricity comes at the cost of farmers displaced from their valleys and salmon prevented from spawning. Expressways benefit commuting car owners but not unemployed residents of inner-city communities. Affecting diverse populations, these technologies are fundamentally different from, for example, medical treatments whose benefits and side effects apply to the same individual.

Devaluing Nature and Culture

Both narrow and short-term problem-solving cause specific harms beyond their selective attention. One of the most recently identified concerns about technological hubris is the neglect of natural environments. By prioritizing technological solutions, designers inevitably focus attention on intended technical outcomes rather than on the unintended consequences and externalities of engineering. Many of the historical and ongoing cases of technological fixing discussed here (DDT, hydrocarbon fuels, plastics, nuclear waste, hydropower) have created identifiable environmental harms. It is safe to say that all human technologies trace back to varying degrees of environmental detriment either in their use of raw materials, manufacture, distribution, usage, or ultimate disposal. If that statement sounds exaggerated, I argue that it is because modern societies have consistently overlooked the net effect of their technological choices.[23]

There is also a case to be made for a broader but equally influential effect of technological fixes: the devaluing of nature itself. The case for societal over-enthusiasm and shared myopia about technology's trumping of nature was made in the first phases of the modern environmental movement. As mentioned above, the turn-of-the-century Sierra Club argued in favour of conserving natural wildernesses from encroachment by modern cities. After the Second World War, ecologist and forest manager Aldo Leopold (1887–1948) decried consumerism, which was turning isolated lakesites into tourist destinations and parking lots for car-driving fishers. He criticized the simple-minded technological fixes employed in the National Parks, which culled wolves to ensure that hunters could be guaranteed the opportunity to bag a deer; the rising deer population consequently denuded trees of their bark and destroyed entire forested hillsides.[24]

Leopold noted that the American southwest had been transformed over the previous century by ranchers and over-grazing livestock. This led to progressive deterioration of the plants, soils, and animal community, an outcome that the early American settlers had not anticipated. He argued that similar situations were happening even more frequently in his time. The "dust bowl" catastrophe of the 1930s, when the fertile topsoil of much American prairie pastureland was lost, had not been redressed over a decade later because short-sighted practices of grazing and deep ploughing had not been altered (figure 7.2). Instead, Leopold argued, farmers had over-used new technologies (small tractors and combine harvesters) and focused on immediate production and economic survival for themselves rather than on a sustainable community ethic. The historical evidence appeared to indicate that human-made changes had often been ecologically violent and of a dif-

Figure 7.2. Scientific land usage, TVA literature, 1942.

ferent order than evolutionary changes. This practical fact, he claimed, should make human behaviours much more cautious and long range.

Yet Leopold was not arguing for a rejection of technological solutions wisely conceived. His career was closely linked to commercial realities. From the mid-1920s he was a director of the US Forest Products Laboratory, a government facility researching and developing wood products as a sustainable resource. A decade later, appointed professor of game management at the University of Wisconsin, Leopold's attention shifted from plant to animal life as a sustainable resource to be managed wisely. He suggested that protected wildlife refuges and hunting laws were not adequate to protect ecosystems. Instead, he increasingly argued for landowners and government to responsibly manage habitats in ways that would encourage interdependence of diverse species, including humans themselves. Leopold described how, when industrial productivity became the overriding goal of farming, it led to soil erosion and the undermining of a community ethic. In its place, he called for a "land ethic" that incorporated sensitivity to other species and environments into modern human concerns.

Leopold also warned that the mid-century trend towards large-scale farming was unsustainable. Mega-farms were designed according to an artificially simple biotic pyramid or ecosystem that reduced the complexity of interdependent species, soils, and chemical resources to cows, manufactured feed, and water supply from deep aquifers. Because these systems are heavily reliant on external technological systems – single-breed farm animals, industrially produced feed, and irrigation systems – they are neither self-sufficient nor sustainable. Similarly, he suggests that human manipulation of natural ecosystems can be short-sighted and disastrous, describing how international transportation of farm products short-circuits such cyclical and stable activities.[25]

Rachel Carson's writing, as discussed in chapter 6, reiterated Leopold's concerns. Her research questioned popular faith in technological progress, an attitude that grew alongside increasing awareness of declining natural environments. Coupled to this faith was unfounded confidence in environmental management and trust in technological solutions to environmental problems. Carson's book had been paralleled by the more polemical writings of American political philosopher Murray Bookchin (1921–2006), who analyzed modern technological systems and their environmental and societal harms. His socially infused science, known as social ecology, sought to identify fundamental faults in modern society, building on the perspectives of ecologists like Leopold and Carson but informed by political theory. Bookchin developed notions of "communalism," or non-hierarchical human interactions that would consider the well-being of all species and environments equitably.[26]

For Bookchin, technological fixes were emblematic of dangerously limited modern perspectives. He identified contemporary environmental thinking and attempts to address technological side effects as "vapid environmentalism," claiming that they were dominated by short-termism:

> As our forests disappear due to mindless cutting and increasing acid rain, as the ozone layer thins out because of the widespread use of fluorocarbons, as toxic dumps multiply all over the planet, as highly dangerous, often radioactive pollutants enter into our air, water, and food chains – all, and innumerable other hazards that threaten the integrity of life itself, raise far more basic issues than any that can be resolved by Earth Day clean-ups and faint-hearted changes in existing environmental laws.[27]

A contemporary Norwegian philosopher, Arne Naess (1912–2009), became a similar conduit for radical environmentalism and criticism of technological fixes. He first summarized his views about environmentalism in a 1973 issue of *Inquiry*, a philosophical journal that he had been founded fifteen years earlier. While praising the emergence of ecology as a scientific study, he argued that responses to ecological problems were inadequately conceived and targeted. In the short essay, he contrasted two contemporary approaches for dealing with environmental issues, characterized as "shallow" and "deep," respectively. The "shallow" environmental movement, Naess argued, had "twisted and misused" the message of scientific ecologists. Its major failing was that it wrongly prioritized environmental concerns. He observed that the proponents were focusing on the "fight against pollution and resource depletion" and had the central objective of ensuring "the health and affluence of people in the developed countries."[28] Naess argued that this first wave of environmentalism, although inspired by Rachel Carson's work and the rediscovery of Leopold's writings, was too often diluted and biased towards immediate concerns and human interests. With its superficial and insufficient attentions, Naess suggested, the shallow ecological approach failed to examine environmental questions holistically and, instead, reduced them to specific local harms with immediate consequences to be corrected by short-term (and often technological) tactics. By identifying environmental issues according to utilitarian criteria, he hinted, problem-solvers and their communities tended to focus on the well-being of "the greatest number" in narrow terms, as discussed above: concerns might be limited to company stockholders or residents of a neighbourhood, for example. Such attentions obscure the recognition of larger implications affecting the well-being of other human communities and other life forms, especially those far away in geography or time. In short, Naess notes that "shallow" environmental considerations opt for focusing on human interests first; attending to the local scale; identifying the best environmental solutions as those which will cause the least disruption to the status quo; and encouraging superficial "Band-Aid" solutions to problems.[29]

Naess and Bookchin highlighted how modern attentions have favoured Western cultures and Western environmental solutions. These attitudes, focusing on avoiding nearby harms (NIMBY, Not In My Backyard), often exported problems, and purported solutions, to less developed countries. Thus, Western shipping companies export the ship-dismantling industry to less developed countries, which has been described both as invigorating regional economies and devastating local environments.[30] The familiar financial tactic of identifying some outcomes as beyond the responsibility of the client is thus extended to the distant environmental harms allowed by globalized economies. Errors of omission (failing to protect impoverished communities from Western refuse) are mirrored by errors of commission (the creation of new problems for non-Western cultures). Naïve technological fixes for marginal populations in poor countries such as Malawi have typically included imported diesel generators, which cannot be fuelled without further imports, and solar-powered devices that cannot be manufactured or repaired locally. Both Bookchin and Naess disapproved of the notion of technologies as a shortcut to leapfrog more demanding social change and recommended their replacement by more morally conscious "appropriate technologies," as discussed in chapter 8.

Eroding Social Resilience

Technological fixers address societal problems, large and small, with inventive innovations. Examples have ranged from specific (e.g., airport scanners to deter specific forms of terrorism) to generic (e.g., plastic products to provide consumer contentment). By opting for technological solutions to human problems, decision makers forego other options that can complement, improve, or substitute for technical measures. While this may be effective in some circumstances, the consistent choice of technological fixes can reduce abilities to recognize or apply other methods. The streamlining of problem-solving methods threatens to make alternatives less available. This is a kind of technological momentum that limits deviation and causes societal options to whither. The term "resilience" refers to the ability of a system to maintain itself in the face of a disturbance; it is commonly employed by ecologists to describe the robustness of an ecosystem that encounters environmental perturbations. "Social resilience" is a more recent but similarly broad term to describe how communities or entire societies recover when problems occur.[31] The term usefully identifies a closed system: problems affect ecosystems (or society) and the solutions are sourced from within that system via its inherent properties. This is fundamentally different from the notion of technological fixes, which applies external solutions (new technologies) to a problematic system (e.g., society). This suggests that studies of social resilience are complementary to attempts to devise technological fixes. However, social resilience is much more than Alvin Weinberg's simplistic contrast between wise technologists and "social engineers" attempting to "manipulate social behavior."[32]

Social resilience concerns a wider swathe of actors than merely social scientists and policy-makers. It suggests aspects of society and culture distributed among its members and institutions. For example, responses to a flood or hurricane might include congregating at a corner shop or church as a familiar meeting point, locating victims via group searches, and housing displaced persons in a local school. Such community resilience is an alternative to technological methods of disaster relief, which might involve air-dropped shelters and evacuation by helicopters. Neither option is necessarily better or faster, but if faith in technological solutions becomes dominant, the social capital of communities may be undiscovered or lost.[33]

Short-Circuiting Democracy

Disaster relief illustrates the political environments and assumptions shaping technological choices. A 2014 Red Cross study promoting the use of emerging technologies in disaster relief identified its principal contributors as futurists, technologists, business leaders, donors, governments, and academics, and a subsequent dialogue with a wider range of "stakeholders." Perhaps unsurprisingly, the chosen cohort identified technologies as the key to improving urban resilience. Their sponsor cited "incredible stories of ways technology has helped people move past their worst days":

"Airbnb helped me find shelter."

"The Red Cross app helped me perform CPR."

"A mobile money transfer helped me reopen my small business."

"The ICRC and Google partnered to help me find my missing sister."

Each of these online services prioritized individuals having access to communication technologies rather than the actions of communities or grassroots organizations. As the project's spokesperson noted, smart phones could expedite aid "even where ... local training is lacking or limited," seemingly bypassing social routes and placing sanguine reliance on technical solutions. As a general approach to responding to disasters – which not uncommonly destroy telephone networks and electricity supplies as well as other infrastructure – this suggests a lamentably narrow approach to problem-solving.[34]

The reliance on experts (summarized in the same document as "curated from various sectors," including "technology, business, government, humanitarian/development, philanthropy and academia") carries political implications. This orientation and confidence legitimated the authority of engineering but disfavoured expertise from other disciplines. In the Red Cross study, those experts were working in technological-fix mode, applying their expertise in communication technology as self-assured solutions to acute human problems. In the process, resilience of the target communities was at best supplemented and at worst circumvented as a resource to be tapped.[35]

Such examples suggest insidious side effects of technological confidence, which can not only foster and sustain technical elites but also disempower other social groups. A central concern is that technological fixes relegate responsibility for solutions to selected engineers or planners. This hierarchy of knowledge and power generally endangers weaker parties (e.g., relatively voiceless human populations or other species) and reduces their autonomy. It may concentrate power in the hands of favoured social actors (e.g., policymakers, companies, designers) or regions (e.g., affluent countries) rather than permitting all parties to take part in decision making. This pragmatic outcome of placing trust in experts also has implications for individual responsibilities. If technological responses to disaster relief are to be expected, is it then best to wait for the well-equipped aid teams to arrive or, alternatively, to make sure that one has a smart phone and the appropriate apps available? In either case, social resilience fades as a consideration. By contrast, approaches that acknowledge community resilience downplay reliance on individual resources and crucial technologies. This suggests that the promises of technological fixes bypass these interdependencies and provide unreliable self-confidences while defusing collective responsibility.[36]

This chapter, then, identifies key shortcomings of technological fixes: their fostering of naïve trusts and myopic attentions that are short-sighted and selective, and that tend to devalue cultural practices and nature itself. And by undermining other social and political options, technological fixes progressively paint society into a corner. The final chapter asks whether historical and contemporary evidence provides insights into what a future for more sustainable modernity might look like.

8   The Future for Fixing

Taming an Infatuation

The previous chapters have explored the attractions of new technologies conceived as solutions to modern problems. From scientists and engineers to policy-makers, entrepreneurs and wider publics, this seemingly cool and rational conviction has seldom been recognized as being emotionally charged. Yet for each of these relevant social groups, the relationship with technical innovation has been akin to falling in love. New technologies represent irresistible appeals and inspire unquestioning acceptance. Technology's responsiveness to our every immediate need, real or imagined, gains our implicit trust. Only in retrospect may such confidence seem naïve and misguided.

Like in a human relationship, modern society and its technologies have become mutually dependent. The analogy suggests that a lasting relationship may require tempering the torrid love affair. Technologies and the solutions they give us *are* compelling, stimulating, and transformative. But can modern societies avoid the immediate gratifications of technological fixes in favour of stable bonds?

Enduring Faith in Fixes

This book adopts a historical perspective to explore how technological confidence has come to captivate modern society, but the last two chapters track more recent unexpected outcomes that followed enthusiastic adoptions.

Some side effects of major technologies are common knowledge today, but they are seldom investigated beyond their specific cases. For modern publics, technological solutions – particularly those just over the horizon – remain seductive.

There are several interlinked reasons for this continuing popular support: the ongoing promotion of the paradigm by contemporary technological adventurers, channelling the spirit of interwar technocrats and science fiction fans; innovating companies seeking new consumer markets for novel problem-solving products and promising an updated version of the postwar future; and media communications, both by traditional sources and by grassroots participants in social media, echoing and extending the hopes of those players. The economic power of consumerism drives governments, too, to conflate product innovation with social progress, and often with the assumption that technological change can be triggered and prepared for but not resisted. A more diffuse attraction is curiosity about yet unexplored human ambitions and the complementary motivations provided by fears: looming problems that compel reassuring solutions. In short, contemporary culture remains skewed towards technological optimism by the pressure of powerful social actors and their rhetoric of progress. Wrapped within this cozy worldview, the narrower confidence in technological fixes can nestle unquestioned.

For over a century, a handful of compelling preachers have proselytized this shared faith. As discussed earlier, none of them shaped wide publics but each influenced distinct cohorts: Howard Scott for technocrats and early science fiction readers; Richard Meier for postwar academics and development agencies; Alvin Weinberg for American policy-makers and young engineers. There have been several figureheads since then.

Steve Jobs (1955–2011), for example, promoted Apple Inc. as a channel of technological agency for human solutions. The company had captured two markets missed by the largest computer company of the period, which introduced its IBM Personal Computer (PC) for small businesses in 1981. Apple's first success was as a supplier and inspiration for the embryonic amateur computing movement via the company's Apple I (1976) and Apple II (1977) computers. Along with competitors Commodore and Tandy Radio Shack, the company attracted a generation of American computer experimenters who wrote their own software and sometimes extended their computers with sensors and output devices. The Apple II proved popular not just among computer hobbyists but also in university research labs that traditionally had improvised equipment to conduct experiments rather than buying expensive and preconceived off-the-shelf devices. These "homebrew" creations (some, like the Apple I, emerging from the eponymous Homebrew Computer Club in Silicon Valley from the mid-1970s) fitted their users' individualistic needs and encouraged their builders to conceive them as generic problem-solving devices.[1]

The second audience captured by Apple was composed of creators and artists from non-technical disciplines. The Apple Macintosh computer (1983), adapting elements of a point-and-click graphical user interface (GUI) conceived by engineers at the nearby Xerox Palo Alto Research Center (PARC), was touted as easy-to-use by non-programmers. An early television advertisement soothed:

> It's more sophisticated, yet less complicated; it's more powerful, yet less cumbersome; it can store vast amounts of yesterday or tell you what's in store for tomorrow; it can draw a picture, or it can draw conclusions. It's a personal computer from Apple, and it's as easy to use as this [finger and mouse click]. Macintosh: the computer for the rest of us.[2]

Its buyers included writers empowered by word-processing and desk-top publishing, and graphic artists enthused by mouse-directed painting and graphics software.

These two subcultures – computer experimenters on the one hand and creative non-technologists on the other – briefly co-existed via such products but thereafter diverged. The Apple I computer had required savvy users to add a keyboard, power supply, and video monitor, and to program it in BASIC – a challenging set of demands for rank novices. By contrast, the Macintosh computer was notoriously "closed," offering no output ports available to hobbyists to interface it with the outside world. Yet, for both subcultures, Apple spawned zealous supporters who identified the corporation with an ideology of personal liberation, conspicuous consumption, and technical progress.[3]

Apple's origins in the Bay Area south of San Francisco were shared with Stanford University and the burgeoning postwar technical culture of Silicon

Valley, the collection of companies that has incubated generations of electronics engineers, would-be entrepreneurs, and start-up companies. It was also the home of Richard Meier, who focused on technological approaches to problem-solving. This Californian enclave became the centre of popular technological faith in America. By contrast, MIT, its east coast academic counterpart in championing American innovation, had become more visibly associated with military contracts and corporate technologies. Californian products and spokespersons exported their credo of individual enablement internationally through companies such as Intel, Hewlett Packard, Google, Facebook, eBay, and Uber.[4]

Perhaps because of these associations, the Bay Area has also nurtured another constituency having less obvious ties to technological confidence. Arguably, an important trigger for the exploding popularity of the counterculture of the 1960s was the freedom of speech and civil rights confrontations at the University of California, Berkeley, across the Bay from San Francisco, and the varied cultural options enabled by the pattern of population mobility of the west coast.[5] California generally, and the Haight-Ashbury district of San Francisco in particular, became a mecca for those seeking freer lifestyles and alternatives to "the establishment" – power-holders then identified as government, law-enforcers, and corporations promoting conservative politics, peacetime militarization, and traditional social values supported largely by the older generation. In retrospect, this countercultural opposition appears to be directly antithetical to Alvin Weinberg's technological fixes. As head of a national lab responsible for nuclear energy, Weinberg was then at the peak of his influence, advising presidential committees about technological means of waging war and defusing the likelihood of race riots, and beginning to lecture student audiences on engineering as a social tool.

Nevertheless, the counterculture was a diverse and fluid movement. From it came Stewart Brand (1938–), an eclectic writer who had enduring influence in tracking the shifting flavour of shared technological concerns and enthusiasms. He became best known for *The Whole Earth Catalog*, a periodical that appealed to the individualistic and anti-hierarchical spirit of his generation and that he continued to adapt to new media and audiences over two decades.[6] A patchwork quilt of design ideas, manual skills, inspirational reviews, practical philosophies, and commercially available resources, it col-

lected themes that eventually intersected with the home computer movement and early online communications. His Catalog was oriented towards
information-sharing and self-sufficiency with an amalgam of do-it-yourself
resources. It epitomized a countercultural engagement with independent
tinkering. Before there was an internet, Brand and associates promoted the
WELL (Whole Earth 'Lectronic Link, 1985), a dial-up bulletin board system
allowing users to post and receive public messages and communicate via
special-interest groups, amounting to an early implementation of a virtual
community.[7] In later publications, Brand increasingly enmeshed his ideas
with commercial innovation and new technologies. He linked the east and
west coast tech cultures with a journalistic account of the MIT Media Lab,
which he identified as emblematic of work by technologists to *invent the
future* through technology.[8] Although his Catalog had begun with a clear
stance against "government, big business, formal education, church," Stewart
Brand's activities built bridges with each of these interests. For a time, he
advised the governor of California and helped found a business consultancy,
the Global Business Network, during the 1980s; his publications referenced
academic research, particularly that associated with corporate and consumer
interests; and his later environmental writing sought to address metaphysical
themes through the demanding eyes of a technological rationalist. With his
subtext of empowerment, Brand thus helped to proselytize and broaden
technological faith in wider culture – at least west coast American culture.[9]

The Momentum of Confidence

Contemporary technological fixes are also promoted by our cultural attraction to novelty. Consumer culture, first established in North America but
increasingly taken up worldwide since the late twentieth century, has been
conditioned by new products and has made collective expectations of
progress endemic. These cultural confidences have waxed and waned for
specific technologies. Space flight, for example, arguably captures less popular enthusiasm today than it did over the preceding century. Interwar science fiction, followed by Cold War rocket programs and the Space Race,
caused public fervour to peak with the Apollo missions to the moon, but
interest fell with the subsequent Skylab, space shuttle, and international
space station programs.

These heroic initiatives were not portrayed to the general public as technological fixes: they did not aim or claim to solve immanent problems with a neat technical solution. However, they were certainly understood as technological fixes by successive American administrations. The earliest upperatmosphere aircraft and satellites carrying film cameras had been designed to short-circuit a looming political problem: public concerns about a "Missile Gap" with the Soviet Union. These experimental technologies, some unknown to the American public, had the covert purposes of identifying lagging American military progress and inhibiting Soviet dominance. In the same way, the subsequent space race equated technological progress with political ascendency. With the later diplomatic accommodations of nuclear downscaling and rising international cooperation, the promises of the space age began to evaporate.[10]

More mundane contemporary issues nevertheless continue to evoke widespread technological faith and illustrate the enduringly popular appeal of technological novelty. A contemporary version is the promise of the "smart city." Its claims are typical of technological fixes and breathtaking in their aspirations. Smart cities are an open-ended promise, envisaging technologies that will solve the problems of urban life, including traffic, public safety, and social well-being.

Recall the characteristic attributes of a fix: claims about its simplicity and cleverness; its identification as a straightforward and deliberate technical solution to a social, political, economic, or cultural problem; its punctual focus on an immediate and local issue rather than a consideration of existing systems, social infrastructure, and human constraints; its identification of promised beneficiaries but neglect of potential harms and externalities; its requirement of expert implementation and citizen acceptance.

The rhetoric of smart cities champions a variety of fixes for identified problems but has been sustained by enthusiasts seeking to implement their favoured technologies. "Smart" is loosely defined but hints at the assumed social benefits of information. The intended benefits are often disturbingly vague. It may mean an innovation that is either economically sustainable, resource-efficient, or responsive to the needs of city-dwellers. A common theme is technology to acquire and report urban conditions for stakeholders, commonly defined as local government, businesses, or mobile citizens. The most frequently discussed genre of solutions is integrated information

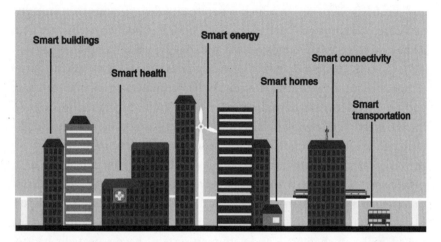

Figure 8.1 Unassailable smart cities: reduced to simplistic rhetoric and imagery, what's not to like?

and communications technologies. Smart cities would add a network of sensing devices, communication links, and control software to new or existing city developments. Integrated systems could predict congestion of roadways, public transport, and even sidewalks and take corrective actions. High-efficiency street lights could illuminate a pedestrian's path and turn off to save energy after they had passed; mobile devices could display optimal routes for pedestrians and identify nearby conveniences; fine-grained environmental conditions – smog, ultraviolet exposure, and weather – could be reported, and avoided, at street level. The subtext of smart cities, like previous technological fixes, is that their innovations can be implemented with a net *saving* of human labour and complexity. Cities such as London, for instance, currently employ automatic cameras and ticketing systems to detect drivers entering the city and imposing "congestion charges," a system that aims to avoid multiplying the bureaucracy of traffic police.[11]

Integrating such useful innovations nevertheless relies on optimism about engineering design at the system-scale. Urban environments are sociotechnical ecosystems that evolve under the influence of disorganized forces; a genuinely smart city would have to be adaptable in the same way. More worryingly, the common experience of city management is generally poor. Consider the management of roadworks to locate and replace water and electricity services, in which separate city departments may scarcely be aware of each other's activities. Such interdependent systems, created ad hoc over decades, tend to co-exist and piggy-back rather than being designed to cooperate productively. Of even greater concern is the maintenance of infrastructure. By contrast to some well-maintained cities, interstate rail and freeway networks have declined; bridges built during the 1930s have increasingly failed owing to lack of monitoring, maintenance, and adaptation to traffic requirements, a key requirement of complex systems. At the city scale, unlike at the interstate level, there may be convergence of public attention, government responsiveness, and taxation mechanisms that sustain such visible networks, but these are essential social and political dimensions of the technical systems they sustain. As illustrated by the examples in chapters 6 and 7, human systems have repeatedly had to adapt to support clever fixes that were initially portrayed as wholly technological and straightforward.

Another dimension of concern, as with earlier technological fixes, is the intended beneficiaries. A smart city could be optimized to promote the interests of drivers (e.g., by temporarily opening routes to make pedestrian or cycling traffic easier) or, quite distinctly, could promote the interests of civic government (e.g., identifying and containing disruptive protesters). Such competing interests have long been recognized. The overtly political dimensions of city governance were incorporated in the design of the wide radial arteries of Paris, the low-clearance parkways of New York State, and the wall that divided East and West Berlin, and can be expected to be more readily configurable in smart cities.[12]

In common with other technological fixes, the promises of smart cities rely on enthusiastic promoters. A technology enthusiast who has promoted both space flight and urban solutions – attracting investors and ardent fans in the process – is Elon Musk (1971–). As an entrepreneur and technology promoter, Musk founded Space Exploration Technologies (SpaceX, 2002), an American firm developing space vehicles as a commercial competitor to organizations such as NASA and the European Space Agency. The company's vehicles gained attention for extending technical capabilities (notably vertical soft landings of a space vehicle on land and on a seaborne platform) and earned revenue by transporting payloads for government and commercial customers.

Musk has also been noteworthy as co-founder of the electric car company Tesla Inc. He combined his two largest technology-demonstration projects by launching one of his cars on a SpaceX vehicle in 2018, making them the first two privately funded products to leave Earth's orbit. Via such publicity coups, Musk is a notable proselytizer of technological faith. He has forecast private space flights for eventual settlement on Mars and potentially terraforming it for humans (a more extravagant promise than mere geoengineering of the Earth's climate, described below) and imagines brain-computer interfaces to liberate human capabilities. He has promoted more down-to-earth conceptual solutions to urban problems, including high-speed transportation between cities via a passenger-carrying pod in an evacuated tube (a scheme dubbed the "Hyperloop") and networks of underground tunnels for moving cars and people. Technological solutions, he claims, are generally best, but his hubris raises controversy along with his public profile. The Hyperloop, imagined as a ground-transport replacement or supplement for aircraft, airports, and railways, has been criticized as a sparse technical sketch that ignores complex human systems and their implications.[13]

As detailed in previous chapters, entrepreneurial caution is generally a weak complement to enthusiasm and profit. Elon Musk and other contemporary technology adventurers spearhead popular enthusiasms and merge them with corporate aspirations by stimulating consumer anticipation and investor confidence. Other high-profile investors and advocates with varying degrees of technological hubris include businessman Richard Branson (1950–), CEO of Virgin Galactic; Microsoft co-founder Paul Allen (1953–2018) and executive Charles Simonyi (1948–); Amazon CEO Jeff Bezos (1964–); and Google executives Eric Schmidt (1955–), Ram Shriram (1956–), and Larry Page (1973–). Such speculation about private and corporate exploitation of the solar system threatens another century of unanticipated consequences. There is not yet national legislation to monitor and supervise such activities, a requirement of the Outer Space Treaty agreed between the USA, the UK, and the Soviet Union in 1967, and now subscribed to by 107 countries.[14]

Such visions, whether urban or extraterrestrial, flourish in a context of entrepreneurism and technological enthusiasm. Confidence in technological fixes becomes implicit: technology as a solution to as yet unidentified human problems is tacitly assumed. The rhetoric of progress shared across media by confident evangelists can encourage public acceptance as the default outcome.

Men Like Gods: Imagining Global Repairs

The scale of attention and infectious self-confidence displayed by space entrepreneurs is characteristic of other past and contemporary technological optimists and fixers. When H.G. Wells wrote his utopian novel *Men Like Gods*, he imagined a technological world that espoused collective wisdom without political hierarchy. Unlimited knowledge and liberty – free discussion, free movement, and privacy – combined to create a rational communitarian society.[15] Yet this sensitivity to both social and technological dimensions was unusual. Others have found such clarity of shared purpose either to be an unlikely outcome, on the one hand, or a trivially obvious benefit, on the other.

Social anthropologist Edmund Leach explored similar themes in his Reith Lectures broadcast on BBC radio in 1968, which open with a provocative statement:

> Men have become like gods. Isn't it about time that we understood our divinity? Science offers us total mastery over our environment and over our destiny, yet instead of rejoicing we feel deeply afraid. Why should this be? How might these fears be resolved?

Leach noted that "we love our machines ... Technical wizardry is just what makes life worth living, it is the badge of civilisation." He noted that the natural and social sciences revealed regularities and insights, and consequently scientists and technologists could no longer be detached: they had a responsibility to apply their knowledge, while meticulously maintaining their connections with both nature and culture. His explorations were not prescriptive but, rather, suggested that science provided powers that had to be patiently absorbed and cautiously applied – hardly the practices of the modern world.[16]

The same year, Stewart Brand's first *Whole Earth Catalog*, subtitled *Access to Tools*, reduced these musings to a strapline and reoriented it as a motivation for his technological compendium:

We are as gods and might as well get good at it ... A realm of intimate, personal power is developing – power of the individual to conduct his own education, find his own inspiration, shape his own environment, and share his adventure with whoever is interested. Tools that aid this process are sought and promoted in the WHOLE EARTH CATALOG.[17]

The god-like power of adventurous technology was a theme familiar to professional technologists, too, and championed by Brand in his later writing. Moving from periodical editor to technology journalist to business consultant, Stewart Brand increasingly adopted the role of futurist and proselytizer. Founded in 1996, his Long Now Foundation aimed to encourage dialogue about long-term thinking, and his book *Whole Earth Discipline*, written a decade later, provides an early twenty-first-century take on technological fixing into the deep future. The book is unabashedly optimistic about grand technological schemes as the only means of ensuring societal survival in the long term. It mutates his old strapline to "We are as gods and HAVE to get good at it." Brand describes his orientation as "scientific rigor, geoeconomic perspective, and an engineer's bias, which sees everything in terms of solvable design problems," a stance evoking Richard Meier's focus a half-century earlier (chapter 3) but with Alvin Weinberg's looser commitment to detail.[18]

Brand identifies moralistic and rebellious environmentalists as his principal opponents and pragmatic technologists as allies: "Engineers are arriving who see any environmental problem neither as a romantic tragedy nor as a scientific puzzle but simply as something to fix."[19] The most ambitious but urgent of these hubristic fixes, he argues, is geoengineering. The pace of anthropogenic climate change is now so rapid, and so unlikely to be managed by conventional human approaches, that the global climate must be ameliorated "by adjusting the nature of the planet itself through large-scale geoengineering." This is a classic Weinbergian technological fix but one of unprecedented proportions. Understanding the fundamental nature of climate change, he suggests, is the wrong focus: we should concentrate instead either on ways of merely controlling the planet's overall temperature or of limiting carbon dioxide emissions, prescribing a kind of global aspirin or antacid remedy instead of adoption of a healthier lifestyle. Alvin Weinberg himself had suggested the much more restrained, but still consequential, technological fix of wholesale adoption of nuclear power to reduce carbon dioxide emissions as a major contribution to climate change.[20]

Ideas for geoengineering fixes have been ingenious but highly contested. Two general options have been proposed: either to concentrate on temperature control by reducing the influx of sunlight or to reduce greenhouse gases, particularly carbon dioxide and methane, which trap solar energy and thus contribute to global temperature rise. Solar radiation could be blocked, for example, by creating a cloudier atmosphere to reflect sunlight away from the earth. Cloud cover could be increased and brightened, for instance, by seeding the atmosphere with sulphur to increase the planetary reflectance (known as albedo enhancement). This is an ironic complement to the unintended pollution of twentieth-century skies by industrial by-products. The sulphur might be distributed via commercial aircraft or shot from cannons, which could produce effects lasting weeks, or seeded in the stratosphere for longer-lasting effects. Alternatively, marine clouds could be increased by unoccupied ocean-going ships, controlled by satellite and powered by the wind, spraying aerosol-sized droplets of seawater (figure 8.2). Still other options for albedo enhancement include growing crops chosen for their high reflectance or painting the roofs of buildings white.[21]

None of these schemes addresses the continuing rise of carbon dioxide caused by burning petroleum-based fuels, which would carry on making oceans more acidic and harming marine ecosystems. Thus, the other, somewhat deeper, approach to a technological fix for climate change involves controlling the concentration of greenhouse gases. Here, too, tactics are inventive but uncertain. One controversial approach that has been trialled on a small scale is dumping iron particles to be consumed by microscopic marine species such as phytoplankton. This source of nutrients is intended to fertilize the seas, producing a bloom of algae that also absorb airborne carbon dioxide; when they die, the organisms carry their biologically bound carbon deep enough to decay slowly, keeping it trapped for decades or centuries.[22] A seemingly more "natural" alternative is large-scale afforestation, planting millions of trees to absorb some or all of the carbon dioxide produced by human activities. A complement to such biological engineering is Carbon Capture and Storage (ccs), a variety of schemes aiming to absorb, concentrate, and dispose of carbon dioxide. As with other technological fixes, the putative solu-

Figure 8.2 Optimistic repairs for the global greenhouse. Salter concept for aerosol-generating ocean vessel to counteract climate change via cloud whitening and planetary cooling. Enthusiast Stewart Brand notes, "the vessel is incredibly cool looking." Salter, Sortino, and Latham, "Sea-Going Hardware for the Cloud Albedo Method"; Brand, *Whole Earth Discipline*, 285.

tions are seldom conceived as elements of sociotechnical systems that would need to be fitted into conventional human practices. Needless to say, each of these fuzzy options carries consequences – scientific, technological, economic, social, ethical, and cultural. New technological systems would have to be implemented within pre-existing systems comprising national laws, cultural practices, public support, and interdependent economies.[23]

Reporting on such ideas, Brand supports their ambition by citing supporters more often than critics and by buttressing his arguments with science fiction stories that imagine terraforming planets for human benefit.

Yet geoengineering is qualitatively unlike Alvin Weinberg's proposals for global nuclear power. While he suggested that societies could make a transition to nuclear power generation, Weinberg was able to conceive of some of the profound consequences because they were already familiar on a smaller scale: the need for scaled-up uranium prospecting and refinement; the complementary requirement of replacing petrochemical fuels for vehicles with electricity; the need to replicate this transition in more and less affluent countries; wholesale adoption by populations, with predictable outcomes for those less able to adapt; and, not least, the problem of rapidly accumulating radioactive waste requiring its own fix. By contrast, geoengineering schemes offer experimental solutions on a planetary scale without precedents to guide design choices. Unlike terraforming the barren planets of science fiction, geoengineering would be experimenting with the only home that humans have. It would also magnify the problems of inequity. Some solutions will affect unlucky geographical regions, species, or vulnerable human populations (such as those inhabiting river deltas likely to face unpredictable flooding following climate manipulation). Thus, geoengineering suffers from the faults explored in chapter 7, being intrinsically overconfident, short-sighted, risky, and unjust on a scale that technological fixes have not yet been. Independent of the scheme chosen, the manipulation of the physical and biological system of the biosphere – the largest and most complex machine known, to describe it from the perspective of technological fixers – is inherently dangerous. A Royal Society panel drawn mainly from prominent scientists and engineers not surprisingly focused on the science and engineering rather than on the social and ethical dimensions of geoengineering. The members recommended that, while geoengineering deserved careful modelling, the numerous technical uncertainties made it worth considering only as an option of last resort.[24] The putative solution falls back on Weinberg's original notion of the *quick* fix: geoengineering is an emergency measure borne of desperation, in which we have no time to evaluate consequences.

It is worth emphasizing that the proponents of geoengineering have seldom analyzed their schemes holistically. The social, political, and economic side effects are usually compartmentalized as externalities. As with other cases of technological fixes there are other human options. These include actions that are already familiar: legislation to limit or prohibit carbon dioxide production from power plants and vehicles; taxation to discourage environmentally harmful activities or commerce; social initiatives to encourage new behaviours; ethical teachings to alter understandings of collective harm and responsibility; dietary changes to scale down the consumption of meat and milk (ruminant digestion being a significant source of methane). Operating more locally and reversibly than planetary engineering, these tactics may have advantages in trialling reversible options and could be argued to more faithfully ally science, technology, and society than do bold geoengineering schemes.[25]

But a second optimistic technology urged by Brand is biotechnology, especially via genetic modification. His scenario of looming crisis (in this case, the Malthusian crisis of insufficient food supplies as climate change reduces arable land) argues for quick fixes out of necessity, but he emphasizes the positive appeal of optimistic and daring technological solutions.

Genetic engineering is qualitatively different from geoengineering and less readily categorized as a mere technological fix. It is a scientific field that has progressed over decades of international research. By contrast, geoengineering arguably represents a spectrum of half-baked technologies cobbled together from the worst traditions of engineering repair. Genetic engineering has become positively associated with human health, which often carries the seal of popular approval for new technologies. For example, Dolly the sheep (1996–2003), now stuffed and on display at the National Museum of Scotland in Edinburgh near where she was raised, represented an aspirational use of genetic engineering. Although Dolly was the first mammal cloned from mature body cells, a procedure still deemed abhorrent and internationally prohibited for humans, the longer-term goal was to develop techniques for genetically modifying animals to produce milk as medicine. Sheep or cows, for instance, would be modified by incorporating human genes into their DNA to produce important proteins that are lacking in human victims of genetically inherited illnesses such as cystic fibrosis and haemophilia. The redesigned farm animals, conceived as living pharmaceutical factories, would supply such proteins in their milk to treat what researchers hope eventually to be a long list of genetic illnesses. A growing variety of other experimental cloning techniques could potentially be used to correct genetically inherited diseases or to develop other therapies. Since the 1990s, in fact, most insulin production around the world has converted from the former method – extraction of the hormone from cow and pig pancreases, a reliably sourced by-product of the meat industry – to a spectrum of genetically engineered variants manufactured in pharmaceutical factories. Developed in research contexts that typically receive scrutiny by medical and licensing authorities and wider publics, these current and potential technological fixes have often been viewed as comparable in principle to other, more conventional ones such as radiation therapy for cancer or heart-valve replacement surgery: risky for patients but relatively free of potential societal harms.[26]

Nevertheless, Stewart Brand's promotion of genetic engineering considers a more contentious domain: genetically modified (GM) foods and species to provide resources of wider human utility. In the most urgent form of the argument, GM is touted as a technological fix for chronic food shortages that result from climate change. An even more contentious version suggests its application for creating species better adapted to deteriorating environments (the most dramatic of which is the potential modification of the human species, as discussed below).

Genetically modified foods are a large and growing class of products created by a variety of genetic technologies. The best known and most popular have been foods that solve relatively minor and non-urgent problems of food production, transport, and consumer appeal. An early example was the Flavr Savr tomato, approved by the US Food and Drug Administration (FDA) in 1994. The tomato was genetically engineered to improve transport robustness and shelf-life while retaining a natural colour. The product failed commercially not because of identified environmental dangers or consumer fears but because of conventional side effects: no practical saving in harvesting or transport and unattractive taste. It proved to be an unnecessary solution to a non-existent problem. More broadly, genetic modification of consumer foodstuffs and animal stocks has introduced new technologies requiring careful evaluation but seldom qualitatively distinctive improvements. For this domain and its interest groups, the motivation for quick technological fixes is consequently more difficult to defend.[27]

Other GM foods have been developed to improve agricultural yield with the aim of reducing food costs. Genetically modified maize, for example, can be designed to be drought-resistant (e.g., Monsanto DroughtGard™, licensed by China in 2013) or to contain proteins that are toxic for certain insects. Each GM variant carries distinctive implications. For instance, adoption of herbicide-tolerant genetic varieties is practised in conjunction with higher herbicide spraying, and excessive use of these chemicals carries specific consequences for ecosystems. The most recognized side effect is herbicide resistance, as weeds rapidly evolve to develop tolerance and require ever-greater quantities of more toxic herbicides in an unsustainable cycle. These chemical treatments increase environmental pollution and potentially affect animal species.[28]

Similarly, insect-resistant GM hybrids threaten ecosystems in large-scale plantings. The US Environmental Protection Agency (EPA) consequently imposed regulations to require farmers to plant unmodified maize in nearby areas to prevent eradicating the vulnerable target strains of the insect pests entirely, thereby seeking to delay the inevitable rise of GM maize-resistant varieties. This is a typical example of a traditional human fix to counter a technological side effect: a legal correction to control over-enthusiastic use of the technological solution in order to reduce later negative outcomes. As with seemingly innocuous additives to plastics discovered to have problems decades after their introduction, opponents of GM maize challenge overoptimism and complacency about unsuspected side effects. Among possibilities are sensitivity of other species to the proteins in the maize, an important concern given the ubiquity of maize in animal feed and human food products. Another is so-called "gene flow," in which genes from the hybrid may transfer to other species of crop with yet uninvestigated consequences.

Arguments have been made, however, for technological fixes for food supply to benefit more urgent and needy audiences, notably populations starved of adequate and nutritious food. A stronger case closer to Stewart Brand's theme has been made for "golden rice." Conceived as a genetically modified variety of rice designed to synthesize beta-carotene, golden rice aims to counter the lack of dietary vitamin A for deprived populations, much in the way that vitamin-enriched flour was introduced in North America and Britain during the Second World War to relieve vitamin B deficiency. Golden rice is a classic technological fix that seeks to bypass political, social, and cultural routes via a speedy technical innovation. Vitamin A deficiency is associated with inadequate dietary variety and has been treated by aid

agencies via high-concentration oral supplements or injections. The promise by promoters of golden rice is that it will supply a simpler, wholly
technical solution to chronic malnutrition, even though it requires an amount
of production, distribution, and monitoring comparable to that of earlier
supplements. Critics argue that the golden rice solution also avoids addressing deeper socio-economic reasons for poor nutrition and, consequently,
defuses action to remedy more pervasive faults in human systems. They note
that it threatens grassroots solutions such as cooperative cultivation and
distribution of conventional dietary sources of vitamin A such as sweet
potatoes, fruit, and leafy vegetables. As with the cases explored in chapter 6,
proponents have tended to overlook the traditional social and economic
supports needed to make golden rice and other dietary supplements viable:
a network of manufacturing, distribution, and, not least, external funding
and research to make this stop-gap fix more sustainable. Opponents criticize
the potential for the consequent dependence of poor countries on corporate
products supplied for profit. In response, proponents have suggested that
(with government funding or corporate largesse) golden rice could be made
available for free to subsistence farmers, and free licences could be offered
to developing countries, a plan that is nevertheless likely to consolidate the
status of disempowered populations and countries. There are also purely
technical issues with golden rice: current varieties, none yet manufactured
commercially, supply insufficient dietary vitamin A for adequate health, and
it is unclear whether cultivation and consumption could become widespread. Stewart Brand has characterized "anti-genetic engineering environmentalists" attempting to "frighten African nations" as responsible for the
difficulties in promoting golden rice and cites the lack of identified health
problems as sufficient evidence to justify its rapid uptake.[29]

Wider arguments against genetic modification as a routine category of
technological fix centre on the issues of complexity and risk. Biological and
ecological systems are more sophisticated than human-designed systems,
and, consequently, genetically modified organisms should be expected to
evince unanticipated outcomes more frequently or dramatically than human-
made technologies. This is a familiar source of problems in large systems
and calls for meticulous attention at the design stage. The concerns about
risk vary across the type of genetic engineering considered. For example,
those that are contained – such as animal cloning and GM hormone pro-

duction in factory environments – present less environmental risk than does the propagation of GM crops or GM insects in the open environment, where they may potentially interact with other species in unpredictable ways. And unlike human engineering of hardware and software, some biotechnologies can self-propagate and evolve in their new environments, resisting containment and control.[30]

Critics cite the "precautionary principle" as a social and ethical brake on innovation for such engineering systems that can rapidly produce unpredictable effects. It calls on decision makers to establish that a change will be safe before it is implemented. This is the complement of typical legislative regulation, which bans (some) products proven to be harmful. Although there is a regulatory system to assess food and drug safety, such policy mechanisms attend to a relatively narrow range of potential harms. The term entered usage in the 1980s and since then has become common in environmental dialogue, particularly in relation to adverse effects from GM technologies. The principle critically connects innovation, benefit, and risk, expressing broad sensitivities that were uncommon during the twentieth century and that remain largely outside legal frameworks today.[31]

Both geoengineering and non-medical genetic engineering, emphasizing the technological and scientific dimensions of these subjects, have encouraged public deference to their experts as reliable social and cultural guides. A prominent lay representative for this contemporary hubris, Stewart Brand argues that this is appropriate: "environmentalists do worst when they get nervous about where science leads, as they did with genetic engineering." His implication is that science and technology provide the tools for expressing human ideals and should not be constrained or redirected by cultural values, political philosophy, or ethics. Indeed, he suggests that rational decision making must be purely scientific and that science should be depoliticized and utilitarian. Thus, governments should refrain, for example, from adopting policies too hastily about banning plastic bags or rejecting nuclear power because their benefits versus harms can be assessed scientifically, unambiguously, and unromantically. Brand's words echo Howard Scott and the interwar Technocrats: "Instead of yelling 'Stop!,' engineers figure out what the problem is, and then make it go away. They don't have to argue about what is wrong; they show what is right."[32]

More Than Human: Technological Fixes for Our Species

This seat-of-the-pants forecasting plays to our collective optimisms about technological possibilities. Brand suggests that human achievements will outpace our problems:

> How about, say, two hundred years from now? If we and our technology prosper, humanity by then will be unimaginably capable compared to now, with far more interesting things to worry about than some easily detected and treated stray radioactivity somewhere in the landscape … Extrapolate to two thousand years, ten thousand years. The problem doesn't get worse over time, it vanishes over time.[33]

This long-term perspective seems to distinguish it from the typical short-term and short-sighted technological quick fix. It expresses the frontier ethic and science fiction optimism of constant expansion. Humans are the most creative of species, it argues; space travel and colonization of new worlds is merely a continuation of our zeal to discover, conquer, and expand. Exploration and colonization of new worlds is human nature, as is outgrowing old environments. According to this perspective, sustainability is less important than curiosity-seeking and innovation. It hints that the journey, not the ultimate destination, is what matters, and imperfect technological fixes along the way are part of the trip, to be experienced but ultimately left behind.

Yet this complacent long-term vision can be decomposed into numerous individual fixes adopted unreflectively, each with worthy aspirations but shallow short-term attentions. Instead of focusing on immediate social, political, or cultural issues, it urges technological improvements to satisfy distant hopes and dreams for entire nations, for the planet, or even for the human species. Such grand and hazy goals, as suggested by the case of geoengineering, nevertheless carry numerous awkward details that may tend to accumulate, rather than vanish, over time. At risk is intergenerational justice: storing up a legacy for future generations to sort out. But Brand's optimism argues that, from a distance, human progress through technology looks rosy and obvious; his opponents might counter that, averaged over the past century, new technologies have implanted systemic problems having slow gestations and still unexplored consequences.

In the early twenty-first century, the most emblematic of these grand ambitions has attracted various labels, notably "transhumanism." The term was coined as early as 1957 by Julian Huxley (1887–1975) – biologist, first director of UNESCO, and brother of novelist Aldous Huxley. He used it interchangeably with "scientific humanism" to describe the imperative created by "new knowledge amassed in the last hundred years" that has "defined man's responsibility and destiny – to be an agent for the rest of the world in the job of realizing its inherent potentialities as fully as possible." As with proponents of technological fixes, Huxley saw this as a task for a cadre of techno-scientific elites, "a few of us human beings ... appointed managing director of the biggest business of all, the business of evolution."[34] At the heart of the modern expression of transhumanism is the enthusiastic identification of new technologies as dramatic means of altering human capacities. The idea of liberating or unlocking greater "humanness" via technologies is an ironic twist. Transhumanism taps into aspirations for personal health and happiness, the most optimistic and widely accepted domain of technological fixes. Yet it carries the potential not merely of fixes for personal health but also more ethically suspect extensions: the goal of achieving technologically enhanced communities via genetic engineering, bio-technical alterations, or retrofitting.[35]

This technological enhancement, proponents argue, will be transcendent, liberating individuals and empowering collective human progress at an unprecedented pace. Academic philosopher Steve Fuller has labelled the theme of re-engineering our species "Humanity 2.0" and argues that new technological capabilities will inevitably alter our collective notions of what it means to be human. In effect, technological innovations will supersede or revamp religion, philosophy, and human traditions. This driving of human capabilities by technological agents is the technological fix writ large, with the most competent experts as directors.[36]

The notions of transhumanism did not develop de novo but drew on the rising technological faith of the twentieth century. After the Second World War, medical interventions became dramatically more powerful. Cardiac surgery repaired and replumbed the heart and arteries, and experimental heart transplants began two decades later. Replacement of body parts – beginning with corneal transplants from cadavers and kidney transplants between identical twins – became routine, and failing organs were supplemented or replaced: by the 1960s, heart-lung machines during cardiac surgery, dialysis machines for periodic treatments of chronic kidney disease, and wearable heart pacemakers were widely employed.

Surgical alterations also became feasible and increasingly available after the war. Plastic surgery repaired not only injured or congenitally malformed features but also enabled cosmetic enhancements as elective surgery. It is notable that these new powers faced contemporary criticism but gradually became culturally acceptable. Facelifts, breast enhancement, and hair transplants rose in popularity first in southern California (aided by the economic motivation of employment in the entertainment industries) and later in other regions. More recently, consumer technologies of body modification have abounded: liposuction and implants, bariatric surgery, skin abrasion, Botox injections, teeth whitening procedures, and muscle stimulators. These possibilities have revolutionized long human traditions of adopting prosthetic aids such as artificial limbs, dentures, and hairpieces, moving them from external functional or cosmetic additions to permanent elective choices now associated with personal expression and lifestyle. Surely, argue proponents of transhumanism, these powers to improve humans will continue to expand indefinitely.[37]

The theme is hardly new. These new bodily options, rapidly identified as commodities for affluent middle-class consumers, inherited on a personal scale the hopes and dreams of science fiction for humanity as a whole. Similar zeal for improving human physiology and intellect had been explored in the golden age of science fiction through the mid-twentieth century. Isaac Asimov's *I, Robot* series of novels imagined how robots could extend human power while remaining dedicated to human needs; Arthur C. Clark's screenplay for *2001: A Space Odyssey* depicted evolutionary development from apes to super-human intelligences tightly coupled to technological powers. As explored in chapter 5, science fiction and popular technological forecasts later converged. A 1965 *Our New Age* Sunday comic strip by Athelstan Spilhaus promised: "By 2016, man's intelligence and intellect will be able to be increased by drugs and linking human brains directly to computers!"[38]

The theme of intellectually superior intelligences moved more assuredly from science fiction to forecasting with the writings of Vernor Vinge (1944–), an academic computer scientist and science fiction writer. In 1993 he suggested that the rapid progress in computing would lead to a point in the foreseeable future that he labelled the "technological singularity." After this point, forecast by proponents as sometime during the present century, human intellectual capabilities would be superseded by artificial intelligence (AI), threatening to leave humans increasingly far behind. This pessimistic view about being replaced by artificial intelligences is sometimes called "posthumanism." Inventor and entrepreneur Ray Kurzweil argues instead that human and artificial intelligence could merge, inducting an epoch of exponentially increasing abilities and transcendence beyond biological limitations. Kurzweil, in fact, had a track record in inventing devices that could be seen both as contemporary technological fixes and as illustrations of the transhumanist route, commercializing some of the first text-to-speech synthesizers, print-to-speech reading machines for the blind, and commercial speech recognition software. Critics have argued that predictions of superintelligences are as simplistic as the fantasies of the previous generation, which foresaw domed cities and space colonies around the corner: many of the technological forecasts could be pursued with enough resources and collective will, but few of the forecasters attempted to explore how society would co-evolve with them. Neglecting how complex sociotechnical systems are likely to adapt to technological perturbations is a common failing of naïve forecasting and futurism.[39]

For non-optimists, the singularity seems to represent the worst outcome of technological fixes: the dramatically unpredictable and uncontrollable consequence of short-sighted technological innovations. The transhumanist vision displays a hubristic faith in technology as the means of human transcendence, but it is peculiarly myopic about how human societies and individuals would be involved.[40]

For technological optimists, however, the progress towards transhumanism can be charted by contemporary achievements and near-term developments. Among the best-case examples of transformative technologies are electronics and computing, which have improved exponentially in memory capacity and computational speed in recent decades. First identified in 1965 by Gordon Moore, then research and development director of Fairchild Semiconductors and later head of Intel Corporation, the density of transistors on integrated circuits was roughly doubling every couple of years. This empirical technological improvement has been christened "Moore's law," with proponents like Kurzweil suggesting that it is indicative of a wider acceleration of human progress. The promise is not acknowledged by all. Critics note that the historical trend of improvement in computing hardware is slowing and is ultimately limited by physics; that some of the continuing improvements have been driven by narrow technical criteria and industry goal-setting rather than more rapid innovation; and that software tends to bloat and slow computation in the opposite way, and so generates a much more limited net social gain.[41]

A futurist and writer with a longer track record in predicting social effects of technology was Alvin Toffler (1929–2016). As a White House correspondent during the late 1950s and scenario writer for IBM, Xerox, and AT&T during the 1960s, he gained familiarity with the practices of American business, government, and technology firms. His 1970 book *Future Shock* captured growing public concerns about the pace of technological change (much as the Technocrats had done during the Great Depression), and his subsequent book *The Third Wave* a decade later focused on his forecasts at the beginning of the "Information Age." Toffler was a technological determinist, seeing new technologies as causing overwhelming disruptions for which social adaptation was the solution. His ideas about "anticipatory democracy" influenced politicians across party lines, including Democrat Al Gore (1948) and Republican Newt Gingrich (1943–). Toffler's vision permeated the Congressional Clearing House on the Future created in 1976 and later co-chaired by Gore and Gingrich. Toffler's 1995 sequel to *The Third Wave* included an effusive foreword by Gingrich, then speaker of the House of Representatives; and, as vice-president during the Clinton administration from 1993 to 2001, Gore championed government action to adapt US society to the internet. Toffler's ideas were also received positively by the Chinese government from the 1980s.[42]

Yet predicting the future is notoriously inaccurate, and the economic drivers of innovation further complicate forecasting. In a cultural environment primed to expect progress, company investment and academic careers increasingly depend on promises of transformative technologies as much as on actual results. This bias, portraying potential progress while ignoring wider outcomes, reinforces the culture of technological fixes.

The case of nanotechnology, a forthcoming field according to some transhumanists, illustrates the prevailing hyperbole of unrealistic optimism, popular faith, and solutions-in-search-of-a-problem.[43] The field was promoted in part by discovery of a new class of molecules resembling the geodesic domes popularized by Buckminster Fuller and, consequently, named "fullerenes." Investigating these materials opened new directions for scientific research and engineering at the molecular level, or nanoscale. The term "nanotechnology" is consequently a catch-all and has described imagined science fiction scenarios of micro-machines assembled on the atomic scale to reproduce themselves or merely new formulations of powders (such as fullerenes) applied to new problems. Nanomaterials have been touted by rebranded divisions of pharmaceutical companies, the materials industry, and semiconductor firms. Hundreds of start-up firms, funded by industry speculators, have proposed applications such as tissue engineering and regenerative medicine, more readily absorbed drug therapies, and carbon nanotubes instead of silicon for microelectronic devices. A Royal Society of Chemistry publication identified the potential for over 120 diverse applications of the foreseeable future, ranging from medicine to environmental remediation, and to materials that improve on nature. Transhumanists are even more optimistic, arguing that technology integration at the nanoscale is a key element for the improved human of the future. Enhancements might include nanobots providing therapies cell by cell; nanoparticle-strengthened implants to replace or strengthen bone; molecular-scale biosensors to detect and regulate body systems. "Bionics" (biological electronics) might improve human hearing and vision or supplement strength with artificial muscles, as imagined in the television series *The Six Million Dollar Man*. Even more importantly, proponents of nanotechnology hope that eventually it will allow the interconnection of brains to electronics, a development that might allow boosting, re-engineering, or even replacing neuron-based intelligence.[44]

Such bold claims of miniaturizing and revolutionizing all current technological competences launched an unassailable wave of unrealistic optimism. As with the enthusiastic corporate and cultural adoption of plastics a half-century earlier, the wholesale application of nanotechnology makes unanticipated outcomes likely and the precautionary principle relevant.

Figure 8.3 Previewing the future at a conference covering "Biomedical Engineering, Medicine and Pharmaceuticals, Life Sciences, Cardiology, Cancer ... and Nano Cosmetics."

Opponents have challenged the more hubristic claims application by application, arguing that the promise of self-assembling nanotechnological machines represents popular pseudoscience at best and a nightmare vision of out-of-control technological side effects at worst.[45]

Mechanical properties at the nanoscale, for example, are unfamiliar and compel caution. As was the case with the environmental effects of synthetic chemical products, nanoparticles have subtle characteristics but important implications. In air, for instance, they prove to be easily dispersed but difficult to measure and can have harmful effects similar to asbestos; in water, their chemical reactivity and mechanical behaviour depend critically on chemical environment; and cell structures in biological materials may be negatively affected by materials such as metals and titanium dioxide in nanoparticle form. Characteristically, the exciting potential of the new field prompted a boom in investor enthusiasm and government initiatives to surf the wave before such side effects were sought, discovered, and investigated.[46]

The mixed results emerging from researchers deflated the initial enthusiasm. Companies have subsequently redefined their promotional usage of the label to describe even more conventional powdered-state products, blunting both the enthusiasm and concerns of consumers. The wholesale fervour for nanotechnology is a contemporary example of historical technological fixes extended into new territory. In fact, the nascent field preempts traditional fixes by offering solutions to non-existent problems. For transhumanists, this reorientation of human perspectives – from medical repair to medical enhancement, and from species-normal to species-enhanced characteristics – is the essence of their aims: technology would be transformed from an efficient fix for traditional human problem-solving into the basis for an endlessly improvable human existence. The vision transcends even popular science fiction scenarios. The *Star Trek* vision of enlightened humans wisely managing their futuristic technologies is replaced by dreams of enhancement, which bypass outmoded cultures and technical limitations and disregard moral convictions.[47]

The transhumanist perspective, not merely placing technology at the centre of modern society but also identifying it as the basis for redesigning humanity, echoes earlier analogs. Previous chapters chronicle the confidence and enthusiasm of technology promoters through the twentieth century as providers of well-being, while noting that many of their inventions had belatedly negative societal effects. The proselytizers among the technological fixers – notably the American Technocrats and Alvin Weinberg – argued for the power of planned technical innovations to directly address or detour around human problems. By contrast, transhumanism takes a different tack between technology, society, and human values. Its attitude exaggerates twentieth-century confidences. Hubristic and self-defining, it identifies technological powers as deterministic and dedicates little attention to the social consequences of enhanced humans in wider society. Transhumanism attempts to argue that human problems, at least for the privileged cohort of adopters, will evaporate as new technological powers sweep forward. A more direct intellectual genealogy can be traced to eugenics; indeed, Julian Huxley, the first to define the aims of transhumanism, was a leader of the eugenics movement and president of the British Eugenics Society between 1959 and 1962. Emerging in the 1880s, eugenics argued that the human species could be improved by scientifically managing human reproduction. Supporters of eugenics did not seek to enhance humanity beyond a presumed God-given limit, but they sought to prevent the dilution of these "superior" traits by "inferior" inherited characteristics. The pseudoscience became popular across the political spectrum at the turn of the twentieth century and in countries across Europe, Asia, and the Americas. In 1915, for instance, the Panama-Pacific International Exposition in San Francisco included exhibits on eugenics supporting its theme of the advancement of civilization.[48]

Organizations and governments implemented bureaucracies based on eugenic policies. The technical details of intentional selection of suitable parents varied from country to country through the interwar period and beyond. Immigration criteria for the United States were designed to filter particular countries and ethnic groups. A combination of legislation, public health administration, and popular attitudes in several countries caused individuals judged to be mentally deficient or mentally ill to be sterilized. More widely still, individuals with physical disabilities or inherited conditions such as deafness were discouraged from marrying. In each country, public opinion largely supported and deferred to expert views.[49]

The social side effects of eugenics hint at more exaggerated consequences for transhumanism, which seeks to create superior humans but via unspecified selection processes. Neoliberals might suggest that the lucky beneficiaries should gain their privileged access to body form and intelligence via the mechanisms of market supply and demand, and draw upon their rapidly rising purchasing power; progressive transhumanists might suggest that governments would ensure their citizens' rights to enhancement; more nationalistic regimes might identify such enhancement as crucial to international competitiveness. In any case, there would be disparity between haves and have-nots. This would be temporary at best and certain to worsen divisions between affluent and poor countries, or technology-privileged and technology-deprived populations. A more fundamental issue first faced by eugenicists, though, was determining superior and inferior traits. Their definitions tended to be circular and to be blind to social presumptions: the "fittest" were those in the upper echelons of society because their privileged social positions reflected their "superior" breeding. In a similar way, transhumanists may tend to favour like-minded (and like-bodied) individuals, introducing a selection bias and consequent social inequalities. In an imagined future world with powerful genetic engineering technologies, infirmities might be prevented or corrected, leading to a more uniformly able and perhaps widely agreed "superior" population but exacerbating the "inferior" status of those unlucky enough to be deprived of it. Warnings about this morally problematic brave new world ushered in by technological faith is not novel, having been raised in 1932 by Julian Huxley's brother, Aldous.[50]

Such sought technological powers exaggerate the problems of technological fixes discussed in chapter 7. Techno-fixers, eugenicists, and transhumanists adopt narrow perspectives: identifying particular problems, focusing on distinct time scales, and attending to specific audiences. They consistently fail to recognize the entwined human systems through which society operates. Instead, they may trade off social cohesiveness for outcomes favouring other parameters. Experts, generally identified as technological enthusiasts, are judged unproblematically to be the appropriate implementers, adopters, and managers of their schemes, thus short-circuiting democratic participation. It is not difficult to appreciate that this privileged perspective systematically disfavours other human contexts and non-human environments.

Imagined Intentional Futures: Irresponsible Innovation or Redirected Ambitions?

The promises and proselytizers explored in this chapter reveal the ongoing conviction of technological faith in modern society. Retaining a historical perspective, it is reasonable to trace the exploration and critiques of alternative paradigms. Can more responsible innovation replace over-confident steps in the dark when considering new technologies?

Cautious technological innovation has not been popular over the past century. The widely expressed concerns voiced during the Victorian era about the effects of industrialization became less frequent in the Machine Age. However, as explored in chapter 6, some technological choices were recognized belatedly as blunders and prompted more general critiques and alternatives. These analyses argued that adoption of new technologies tended to overlook social and environmental considerations at the design change, neglected negative outcomes, and often overtly traded off side effects in favour of economic interests.

In a culture increasingly attentive to local and measurable improvements, longer-term inadequacies were less noticed. Such short-termism has been attacked, for example, by political scientist Steven Teles, who describes US social policy as a "kludgeocracy." He suggests that policy-makers generally opt for imperfect fixes rather than for fundamental reforms.[51] His neologism has a technological origin, coming from the computing term "kludge," a cobbled-together fudge of software fixes that gets around an immediate problem but more often than not makes the software more difficult to maintain. The growing usage of his term suggests how far technological methods have infiltrated traditional social and political approaches, and how problematic they are. The technique was at the heart of Alvin Weinberg's proposal of technological fixes for government policy-making.

On a broader and more positive scale, analysts outside "the establishment" have critiqued its growing reliance on technological solutions and offered long-term alternatives. Among the most important have been environmental and political philosophers Arne Naess, Murray Bookchin, and Ernst Schumacher.[52] As introduced in chapter 7, Naess discussed two broad configurations of environmental consciousness: the concerns of what he identified as "shallow ecology" and "deep ecology," respectively. His tracing of shallow thinking maps onto the solutions favoured by supporters of engineering fixes. This ad hoc approach remains the most popular engagement with environmental problems and, as might be expected, has collected a random assortment of ready fixes and adopters. Stewart Brand has vaunted so-called "Bright Green" tactics to address environmental problems case by case. Coined by writer Alex Steffen in 2003, the approach asserts that innovative technologies provide the keys to environmental sustainability, provided that political and economic accommodations encourage them.[53]

Naess identified this technology-oriented approach as inherently misguided. Energy-saving appliances, on the one hand, are a great improvement over the wasteful devices of past decades. Yet they may, on the other hand, encourage consumers to continue to buy, and eventually discard and recycle, even more such "labour-saving devices"; we may ask whose labour is being saved. Similarly, the installation of "eco-friendly" light bulbs, or participation in Earth Day events, may encourage individuals to feel that they are positively contributing to sustainability while leaving the preponderance of their lifestyle unquestioned and intact. Naess proposed his "deep ecology" as a more principled and holistic perspective. Technologies, he argued, are as likely to create negative as positive effects, and so the choice of technology must consider its social, cultural, and economic ramifications.[54]

Murray Bookchin challenged these sensitivities and solutions, arguing that Naess's approach identified the appropriate cultural currents but required a more consistent approach that fundamentally reconceived society. Both, nevertheless, had similar criticisms of technology. Some deleterious environmental aspects of technology, they argued, relate to how problems and solutions are framed and addressed: typically, affluent present-day populations are favoured, and other interests are neglected. More pointedly, Naess and Bookchin criticized technological fixes as dangerously seductive: employing technology as a shortcut to bypass deeper social corrections, they noted, makes societal inequities that much harder to eradicate.

Their critiques and solutions map neatly onto the ideas of their contemporary, Ernst Schumacher. A British economist who spent most of his career as economic advisor to the country's National Coal Board, Schumacher spent a period in Burma, an experience that had suggested to him that the distinctive values of human lifestyles could not be reduced to modern Western criteria. His use of the phrase "Buddhist economics" highlighted his view that quality of life in modern societies required a more holistic and spiritual sense of fulfilment. Schumacher's influential book *Small Is Beautiful* argued for this broader perspective on social, environmental, and economic issues. As suggested by the title, he presented the case for rescaling human activities to better serve human and environmental needs. Both modern economics and technological development, he argued, need to be recast. Schumacher identified two flawed models of technology: "the super-technology of the rich," on the one hand, and "the primitive technology of bygone ages, but at the same time much simpler, cheaper and freer," on the other. The first was appealing but also wasteful and poorly distributed. The second was back-breaking and inefficient but readily available. Drawing on the work of Mohandas Gandhi, he defined "intermediate technology" between these two extremes:

> The technology of production by the masses, making use of the best of modern knowledge and experience, is conducive to decentralisation, compatible with the laws of ecology, gentle in its use of scarce resources, and designed to serve the human person instead of making him the servant of machines … One can also call it self-help technology, or democratic or people's technology – a technology to which everybody can gain admittance and which is not reserved to those already rich and powerful.[55]

His perspective mirrored views growing in the counterculture and provided a coherent alternative view of how morally defensible technologies should be conceived and valued. Intermediate in cost, complexity, and sophistication, they would rely on people of intermediate know-how and might consequently trade off these attributes by being of intermediate usefulness rather than high-tech. Schumacher identified key attributes as small scale, small harm, mixed technologies, and design adapted to local circumstances. Examples would include small wind generators like those used on American farms between the wars, which could be repaired or even built from scratch from readily available materials such as wood and wire or equivalent power sources harnessing flowing streams.[56]

The characteristics of appropriate technology, as defined by Schumacher and others, argue that it is adapted to the needs, skills, and resources of its users and environments, and tends to emphasize local autonomy, egalitarianism, and sustainability. A technology ideally adapted to its environment is one that relies on locally available materials and human resources for its design, manufacture, operation, and maintenance. The design is required to be environmentally neutral not just for its users but holistically for all affected parties. These characteristics, he suggested, can serve as goals for guiding wise technological choices.

First, appropriate technologies support local autonomy and self-sufficiency by encouraging local expertise in design, production, and repair. By avoiding reliance on centralized skills and authority, they consequently reduce hierarchies and potential injustices.

Second, such a locally oriented scale encourages responsive and wise innovation. This connection between designers and users crucially distinguishes appropriate technology from Weinberg's notion of the technological fix. Operating on a small scale may make designers of appropriate technologies, who are likely to be the users themselves, more alert to genuine needs and contexts, and to immediate side effects.

Third, appropriate technologies encourage diversity, identified by both Naess and Bookchin as an abstract but valuable principle to be promoted. The concept grows from the scientific principle identified by earlier ecologists such as Aldo Leopold: species diversity tends to produce more resilient ecosystems that can adapt to unexpected perturbations. The idea is also compatible with the notion of technological momentum, which argues that the ferment of nascent technologies offers more adaptiveness to social needs than do mature, large-scale technological systems. There is, though, a counterargument against appropriate technologies: by adapting to suit local context, they are unlikely to benefit from economies of scale and so may prove more expensive to develop and more difficult to maintain consistently.

Fourth, appropriate technologies are likely to be more sustainable in resource usage. By seeking to employ locally sourced materials, they encourage clever innovation and adaptation to suit local contexts. This principle of

Table 8.1
Design considerations informed by critiques of technological fixes

1) Are there implicit assumptions at play? E.g.:
   a) Simplistic identification of the problem (trust in reductionism)
   b) Ease of implementation (unsophisticated planning)
   c) Confidence in likely success (belief in inevitable progress)

2) Are there identifiable interests ("stakeholders") with distinct views or sensitivities regarding the technology? E.g. particular:
   a) Social groups
   b) Species and ecosystems
   c) Natural environments

3) Can the technology under consideration be understood as part of a sociotechnical system? E.g.:
   a) How are manufacture, usage, and recovery linked to other technologies and human systems?
   b) How is the technology linked to existing activities and interests of relevant social groups?
   c) How is it linked to wider environments?

4) Could the technological choice have foreseeable side effects? E.g.:
   a) Technological effects on other parts of the system
   b) Social or cultural implications
   c) Environmental implications

5) Could the technology be implemented cautiously? E.g.:
   a) Could it be made sensitive to different stakeholders?
   b) Could outcomes be monitored adequately?
   c) Would it permit corrections or reversals if necessary?

having a closed-loop system involving production, consumption, and recycling was first identified as a basis for maintaining sound ecosystems by Aldo Leopold and is the basis of lifecycle assessment discussed in chapter 6. Some of these design considerations are summarized in table 8.1.

Slippery Vocabulary and Misleading Practices

The rhetoric of novelty, exploiting neologisms like "technocracy," "smart cities," and "transhumanism," has been influential in shaping cultural acceptance of technological fixes. By contrast, the term "appropriate technology" has declined in usage since its peak in the early 1980s. It became increasingly associated with the perceived focus of Ernst Schumacher and Richard Meier: less-developed countries. The concept's relevance and implications for modern urban life were difficult to communicate to professional engineers and wider publics. Labels and meanings consequentially have mutated. The term "sustainable technology" has grown in popularity since the 1990s to challenge it.[57] The word has been adopted by companies and policy-makers as often as by grassroots environmentalists, sometimes employed as a form of "greenwash" to label restricted examples of "sustainability," as discussed in chapter 6. The transition from "appropriate" to "sustainable" arguably diluted the ethical demands of wise design.

A term seeking to recover part of Schumacher's wider social and moral sense of appropriate technology, however, is "responsible innovation." The label has been used since the 1960s but has been adopted more recently for inter-governmental planning of research policy and implementation, particularly in Europe. A 2013 European Union report, *Responsible Research and Innovation* (RRI), describes responsible innovation as

> The comprehensive approach of proceeding in research and innovation in ways that allow all stakeholders that are involved in the processes of research and innovation at an early stage (A) to obtain relevant knowledge on the consequences of the outcomes of their actions and on the range of options open to them and (B) to effectively evaluate both outcomes and options in terms of societal needs and moral values and (C) to use these considerations (under A and B) as functional requirements for design and development of new research, products and services.[58]

Such definitions appear to place responsibility in the hands of designers and funders (particularly government funders), with no overt mechanisms for public participation. This direction by technical elites echoes the ideas of the Technical Alliance a century ago and of Alvin Weinberg fifty years later.

The ethical norms and responsibilities are also ill-defined. Seeking grounds for negotiation and consensus, the same report suggests that "Standards on RRI that can be adopted *voluntarily ... could* include ... a shared definition of RRI, including principles *like* orientation towards ... gender equality, open access, public engagement *etc.*"[59]

By contrast, Schumacher's focus was on designers, maintainers, and communities, and an important feature was the avoidance of hierarchies of power and governance. Central direction of technological design and choice, he suggested, tends to impose solutions that are not well adapted to local circumstances or to weaker parties. A key difference between "responsible innovation" and "appropriate technology," then, is their respective sensitivity to "softer" human concerns. Appropriate technology may sometimes promote cultural and social traditions rather than business growth, for example; responsible innovation, on the other hand, may more often favour the greater good over regional concerns.

Broad adoption of something like the perspectives of appropriate technologies or responsible innovation may appear unlikely. The promotion of deliberate technological futures by ardent proponents, coupled with our collective appetite for novelty and personal benefit, works against more cautious and systematic consideration. As suggested by contemporary futurology, deeper thinking about sociotechnical systems remains uncommon. Among the key aspects identified in the historical cases examined in this book are the under-appreciated frequency – even regularity – of unintended consequences; the poverty of adequate design consideration of such side effects; and, the inherently political dimensions of technological choice. For a century, the trend in technological fixes and consequent side effects has been their scaling up, thereby increasing the vulnerability of regional and even global environments. The interdisciplinary teams studying anthropogenic climate change label it a "wicked problem," in the sense of not being amenable to solution by a single discipline or approach.[60] This growing consensus surrounding the human problems associated with climate change suggests that technological fixes, when they work at all, address problems only at relatively modest scale and in the short term.

As I have sought to show, the history and momentum of technological faith is unsettling. On the one hand, modern culture has become primed to expect and welcome new technological solutions and to disregard critical assessment until they have been widely adopted and found wanting. On the other hand, local incidents have made publics painfully aware of unplanned side effects of specific technologies.

The most consistent thread through this century-long history of techno-logical hubris is the enduring role of rhetoric. Verbal persuasion and imagery have been tools to legitimize optimistic expectations that rely on inadequate evidence or extravagant claims. As the technocrats employed them, speeches in the form of simple tales won over audiences. Early science fiction and pop-ular science magazines portrayed escapist adventures and uncritical futures provided by sage designers. Weinberg's essays and speeches introduced sim-ilarly evocative (and evasive) examples. Contemporary entrepreneurs and technological adventurers carry on the tradition. Just as importantly, graphic illustrations have reiterated the rhetoric of technological solutions for our social world, from the technocratic postcards to lurid covers of *Popular Mechanix*, to mid-century corporate advertising and contemporary online media promoting nanotechnology, smart cities, and geoengineering quests. Recognizing the potency of such imagery, I have anticipated reader predilec-tions by avoiding pictures of side effects (e.g., waste dumps or sea life stran-gled by plastic) likely to be interpreted as partisan or polemical in favour of the unrealistic promises of positive human futures. My aim has nevertheless been to communicate the faith-like nature of such technological assurances: the brief parables, sermons, and catechisms on which they were based; the modern zealots proselytizing the planned utopian future; evangelists for miraculous technological cures; and the liturgies of modern public discourse. At the heart of the analogy is the nature of faith itself: the quality of unrea-soning trust detached from understanding or justification. The irony is that the history of this technological belief has been so poorly supported by ra-tional underpinnings.

Notes

ACKNOWLEDGMENTS

1 Johnston, "Technological Parables and Iconic Imagery"; "Technological Fixes as Social Cure-All"; "Vaunting the Independent Amateur"; "Alvin Weinberg and the Promotion of the Technological Fix." Conference presentations included the Society for the History of Technology (Albuquerque 2015), European Society for the History of Science (Prague 2016), and the joint meeting of the History of Science Society/British Society for the History of Science/Canadian Society for the History and Philosophy of Science (Edmonton 2016).

CHAPTER ONE

1 Weinberg, "Will Technology Replace Social Engineering?"
2 For example, Brand, *Media Lab*.
3 For business leaders more specifically, "disruption" carries similar connotations – the notion that shaking up systems generally produces positive opportunities and beneficial reconfigurations for companies and, by implication, wider society.
4 https://books.google.com/ngrams; Michel et al., "Quantitative Analysis of Culture."
5 Historian Harry Armytage (1915–1998) discussed how such beliefs were expressed in technocracy and science fiction, both explored further in this book. See Armytage, *Rise of the Technocrats*; *Yesterday's Tomorrows*.
6 A well-known claim is that the invention of the stirrup was responsible for the rise of feudalism in the eighth century AD because it allowed for rapid

armed warfare on horseback. See White Jr, *Medieval Technology and Social Change*. The enabling of effective cavalry, he argues, led inexorably to a warrior class and, with it, a system of royal land grants in exchange for military service, with the land worked by subordinate tenant serfs. The underlying claim of "technological determinism" has been criticized by subsequent historians, who argue that the stirrup may well have made new forms of military might and political power possible but not inevitable.

7 Arguably, the recent trend towards development of autonomous weapons systems reifies the concept of military technologies usurping human combat and tactical choices.

8 Historians and analysts of technology conventionally define "technology" as not merely hardware (and, more recently, software) but also as practical knowledge and skills embedded in human-made "systems" of organization and control. Many of them would recognize "sociotechnical systems" as the fundamental object of their study. Such systems are made up of hardware (e.g., mobile devices and telephone networks), embedded knowledge and skills (e.g., apps, operating protocols, and conventions of usage), and human participants (e.g., designers, fabricators, maintainers, users, and non-users). The English language term "technology" crudely combines these aspects, but other languages may distinguish them. Modern French, for example, employs the term *technologie* to mirror English usage describing artefacts but reserves a broad definition for *technique* as comprising skills and procedure, including organizational and management systems; ancient Greek, by contrast, distinguished *epistêmê* (theoretical knowledge) from *technê* (art, craft, application). For most proponents of technological fixes, however, attention focuses on human-made artefacts (prioritizing technical, rather than sociotechnical, perspectives), and the social dimensions may be understood simplistically in terms of *beneficiaries* or *societal outcomes* of the technology.

9 An excellent overview of the analytical science, anthropology, and sociology of prehistory and their interdisciplinary connections is Roberts, *Tamed*.

10 The direction of influence is a key factor in assessing how societies act and react in relation to technological options. Traditionally, it has been argued that populations rose as a *result* of reliable food production (i.e., that agricultural technology enabled human expansion), but more recently it has been suggested that population pressures may have encouraged exploration of new food sources. See Weitzel and Codding, "Population Growth."

11 According to Drucker, "Technological Revolution," the rise of "irrigation

civilizations" founded on technologies of water control led inexorably to a revolution in the social organization of cities.

12 Finch, "Ancient Origins of Prosthetic Medicine"; Thurston, "Paré and Prosthetics."

13 Miles, "Disability in an Eastern Religious Context."

14 Tobias, "Technology and Disability."

15 McKellar, "Artificial Hearts."

16 On the enrichment of staple foods with vitamins, see Ackerman, "Nutritional Enrichment of Flour and Bread."

17 The sociotechnical system of preserving, transporting, and consuming frozen foods, for example, was largely a post-Second World War development involving new technologies (notably refrigeration and microwave-cooking) co-evolving with social and cultural changes (e.g., a declining proportion of primary home-makers and the rise of convenience foods). See Mallet, *Frozen Food Technology*. Technological aids included a rapidly expanding variety of over-the-counter products to increase metabolism, reduce appetite or fat absorption, and exercise machines to burn calories. See Maguire and Haslam, *Obesity Epidemic*.

18 Augustine, "Obesity and the Technological Fix."

19 For example, Chapelle, "Bioremediation of Petroleum Hydrocarbon-Contaminated Ground Water"; Gabrys, "Plastic and the Work of the Biodegradable,"; Royal Society, "Geoengineering the Climate."

20 Engineering professions adapted to the contemporary environment of terrorist threats by creating special interest groups to promote security technologies and funding for technological fixes. Among the first was the Homeland Security group of SPIE, the optical engineering society, which aimed to "stimulate and focus the optics and photonics technology community's contributions to enhance the safety, counter homeland threats, and improve the sense of well being." SPIE, http://www.spie.org/Announcements/index.html#homeland, viewed 14 May 2003, page no longer extant.

21 Usage of the term "solutionism" initially referred to "confidence that solutions can be found" and described a generic optimism related to American international and economic policy. It has been more recently appropriated to label the philosophical (or at least marketing) view of Silicon Valley entrepreneurs. See Morozov, *To Save Everything, Click Here*. As argued in chapter 5, such groups are, in fact, the inheritors of a longer and more nuanced tradition.

22 Themes such as "technology-enhanced learning" and "technology-mediated communication" sustain growing industries, and the Apple slogan "there's an app for that" offers software solutions for putative human needs.

23 The historical period since the mid- to late nineteenth century is often labelled "late modernity." Although the time scale and attributes have been variously defined, some broadly agreed characteristics include: technological systems of communication and transport (e.g., telegraph and railways); modern structures of governance; cultural engagement with industrialization; global interactions as an intrinsic feature of economics and communication; and professionalization of scientific and engineering professions.

24 A comparably blind faith in the powers of scientific knowledge and methods, known as "scientism," is today widely derided. The term "technologism" has occasionally been used to describe the faith in fixes that are the theme of this book.

25 Prior accounts of American technocracy have generally focused on the organization's rise in popularity and fall from attention before the Second World War, neglecting interactions and influences later in the century. Jordan's excellent account of industrial ideology centres on the interwar period but does not focus on Technocracy and its particularly radical claims about technology as social cure. See Jordan, *Machine-Age Ideology*. Similarly, most scholarly attention to Weinberg has concerned his role in nuclear engineering and policy (e.g., Johnston, *Neutron's Children*).

26 The principal primary sources are the Technocracy Fonds at the University of Alberta Archives and the University of British Columbia Archives (henceforth UAA and UBCA, respectively), rich in late twentieth-century documents from Technocracy regional chapters and deposited up to the early 2000s; Weinberg's self-classified papers, divided between the University of Tennessee Modern Political Archives, Baker Center for Public Policy, Knoxville (MPA.0332) and the Children's Museum of Oak Ridge, Tennessee (henceforth MPA and CMOR, respectively), with the latter first made available in 2016. Richard Meier's papers were first made available at the Bancroft Library, University of California, Berkeley, in 2018.

27 On discourses directed towards experts, see, for example, Overington, "Scientific Community as Audience"; and Winsor, *Writing Like an Engineer*.

28 See Wilson, Pilgrim, and Tashjian, *Machine Age in America*; Banham, *Theory and Design in the First Machine Age*; Hughes, *American Genesis*.

29 Technical Alliance, "Technical Alliance," 1. See also Armytage, *Rise of the Technocrats*.

30 An opposing theoretical view is social construction of technology, which argues for human agency in shaping innovation and, with it, our social world. These competing themes are explored throughout this book.

31 Russell, *Religion of the Machine Age*. See also Noble, *Religion of Technology*.

32 Jay N. "Ding" Darling (1876–1962), *New York Herald Tribune*, published the day after the atomic bombing of Hiroshima. Darling's editorial cartoons often highlighted the quandary of scientific progress alongside enduring social realities. From Alvin Weinberg's personal collection.

33 For example, Meier, "Automatic and Economic Development."

34 For example, Meier, "Hopeful Development Path for Africa."

35 Del Sesto, "Wasn't the Future of Nuclear Energy Wonderful?"

36 "Weinberg Touts 'Technological Fixes' to Stabilize World," *Oak Ridger*, 12 October 1966.

37 For a nuanced account of the socio-political context of spaceflight, see McDougall, *Heavens and the Earth*.

38 For relatively sparse mid-century critical assessments, see Mumford, *Technics and Civilization*; Ellul, *Technological Society*; Marcuse, *One-Dimensional Man*.

39 Leslie, *Cold War and American Science*.

40 Carson, *Silent Spring*.

41 For example, Love Canal, near Niagara Falls, New York, became infamous during the late 1970s as a community ruined by leakage from a toxic waste dump. Technological opposition in America was directed successively towards the Supersonic Transport initiative, the Alaska Oil Pipeline and – particularly after the Three Mile Island incident in 1977 – nuclear power plants. International incidents included spillages from the oil tankers *Amoco Cadiz* (1978) and *Atlantic Empress* (1979). Later incidents, such as the *Exxon Valdez* (1989) and *Deep Water Horizon* (2010), fuelled public debate about societal reliance on large-scale technological systems, ironically while promoting technological fixes for cleaning up after such accidents. See, for example, Horwitch, *Clipped Wings*; Coates, *Trans-Alaska Pipeline Controversy*.

42 Weinberg, "Nuclear Power and Public Perception," 279.

43 Burns and Studer, "Reply to Alvin M. Weinberg."

44 Franks, Hanscomb, and Johnston, *Environmental Ethics and Behavioural Change*.

45 Naess, "Shallow and the Deep."

46 Drengson, "The Sacred and the Limits of the Technological Fix."

47 Schumacher, *Small Is Beautiful.*

CHAPTER TWO

1   Wood, "Birth of the Technical Alliance," 16.

2   Streetcar safety arguably was driven not by responsible innovation but by the pressure of financial losses to transport companies to compensate injured passengers and by impending government-mandated design changes. "Open platforms gave way to platforms with gates, gates to fully enclosed platforms. Faced with the threat of government action, companies took pre-emptive action." See Welke, *Recasting American Liberty*, 30–1. Patents included Rowntree and Spencer, "Combined Street-Car Pneumatic Door Device"; Beck, "Steps for Railway and Street Cars."

3   Similar designs of the period included the Peter Witt trolley and J.G. Brill streetcar designs circa 1916. See Middleton, *Time of the Trolley*. Engineering rationalism was nevertheless not deterministic: older open platform designs such as the cable-drawn cars used in San Francisco co-existed with the safety designs adopted in other cities, and horse-drawn trolleys worked alongside electric streetcars in New York through the 1920s.

4   Wood, "Birth of the Technical Alliance," 17. Representing the spirit of the times, the term "technocracy" was coined independently by several individuals around 1919. See Smyth, "Letter to the Editor." The term also gained currency in other countries, notably interwar Germany. See, for example, Lenk, *Technokratie Als Ideologie.*

5   Mead, *Blackberry Winter*, 195–8 (emphasis added).

6   Hubbert and Doel, "Oral History Interview," American Institute of Physics.

7   Wood, "Birth of the Technical Alliance," 15.

8   For a well-rounded contemporary journalistic investigation of Scott's career, claims, and philosophy, see Raymond, *What Is Technocracy?* and, particularly, chapter 3, "Who Is Howard Scott?," 100–19.

9   Akin, *Technocracy and the American Dream.* For a supporter's account, see Parrish, *Outline of Technocracy.*

10 Scott, "Origins of Technical Alliance and Technocracy." The greedy business speculation, political corruption, and consequent social disparities during the decades that followed the Civil War were first labelled and satirized in Twain and Warner, *Gilded Age.*

11 Veblen, *Theory of the Leisure Class.*

12 "Any such innovation that fits workably into the technological scheme ... will make its way into general and imperative use, regardless of whether its net ulterior effect is an increase or diminution of material comfort or industrial efficiency." See Veblen, *Instinct of Workmanship,* 314; Bimber, "Karl Marx."

13 Chase, *Tragedy of Waste.* See also Westbrok, "Tribune of the Technostructure," who describes him as a "technocratic progressive."

14 On Taylorism, see Kanigel, *One Best Way.*

15 Jordan, *Machine-Age Ideology;* Segal, *Technological Utopianism in American Culture;* Brick, *Transcending Capitalism.* By contrast, Edwin Layton argues in his revised 1971 book that few American engineers during the "progressive era" were radicals. See Layton, *Revolt of the Engineers.* On varied disciplinary perspectives, see Misa, Brey, and Feeberg, *Modernity and Technology;* and Hard and Jamison, *Intellectual Appropriation of Technology,* 212–14.

16 Soddy, *Wealth, Virtual Wealth and Debt.* On Soddy's interest in societal questions, see Sclove, "From Alchemy to Atomic War."

17 Soddy, *Role of Money,* 4, 7. The "erg" is a physical unit of energy defined in the centimetre-gram-second (cgs) system of units.

18 On the "hedonic calculus," or computations to optimize the greatest happiness for the greatest number, see Bentham, *Introduction to the Principles of Morals and Legislation;* Mill, *Utilitarianism.*

19 For a summary by his contemporary, J.S. Mill, see Mill, *Auguste Comte and Positivism;* Donnelly, *Adolphe Quetelet.*

20 Geddes, "Analysis of the Principles of Economics." The paper built on the work of Scottish physicist Peter Tait, who had been an important contributor to the science of thermodynamics. Soddy himself worked at the University of Glasgow and University of Aberdeen from 1904 to 1919.

21 Gorelik, "Bogdanov's Tektology." Bogdanov's ideas can be traced in early cybernetics and science fiction, discussed in chapters 3 and 5, respectively.

22 Sorokin, *Contemporary Sociological Theories,* chap. 1; Mirowski, *More Heat Than Light.*

23 In the young Soviet Union, the large Vkhutemas technical school (1920–30) was a close analogue of Bauhaus. See Rodchenko, *Experiments for the Future;* Lodder, *Russian Constructivism.*

24 Schwartz, *Werkbund;* Banham, *Theory and Design in the First Machine Age.* Similarly, a Russian "cult of technology" contrasted the technologically

progressive Soviet state with its reactionary Tsarist predecessor. See Joseph-son, *Would Trotsky Wear a Bluetooth?*; Johnson, "Technology's Cutting Edge."

25 Technical Alliance, "Technical Alliance." 1.

26 Veblen, *Engineers and the Price System*; "Memorandum on a Practical Soviet of Technicians." See also Dorfman, *Thorstein Veblen and His America*; and Tilman, *Thorstein Veblen and His Critics.*

27 Two articles appeared in the IWW periodical: Scott, "Scourge of Politics in a Land of Manna"; "Political Schemes in Industry."

28 Rieger, *Technology and the Culture of Modernity*; Edgerton, *Shock of the Old*; Nye, *Electrifying America*; Daunton and Rieger, *Meanings of Modernity.*

29 For example, Wilson, Pilgrim, and Tashjian, *Machine Age in America*; Banham, *Theory and Design in the First Machine Age.*

30 Scott, "Newspaper Interview," 15 (emphasis added).

31 The most thorough account of this period is Elsner Jr, "Messianic Scientism." On the later history of the organization, see Adair, "Technocrats 1919–1967."

32 Scott, "Radio Address," https://www.youtube.com/watch?v=xyORanhrXBQ.

33 For example, "What Is Technocracy?"

34 Technocracy made inroads in British Columbia and Alberta, where entrenched Social Credit governments offered a populist philosophy promising economic restructuring, but it was just as difficult to place on the left-right political spectrum. The overlapping memberships are suggested by Joshua N. Haldeman, research director of the Canadian branch of Technocracy Inc. (1936–41) and provincial coordinator of the Saskatchewan Social Credit Party (1943–50). See Keating Jr, "Joshua N. Haldeman"; Adair, "Technocrats," 25.

35 Scott, "Public Lecture." On technology-related accidents, which became an important rhetorical element for Scott, see Burnham, *Accident Prone.*

36 Scott, "Birthday Talk." The replacement of overt force by calm rationalism in redirecting behaviours is also characteristic of the claims of behavioural psychologists between and after the wars. B.F. Skinner's (1904–1990) "radical behaviourism" was then current, informing his novel *Walden Two* and his subsequent book *Science and Human Behavior*. The technocratic and behaviourist devotion to quantification and rejection of psychological interpretation have evident links with logical positivism, which was also then at its zenith for American philosophers of science.

37 Scott, "Design, Direction or Disaster."

38 These trends concern the central claims of Technocracy: the rise of production, consumption, and waste; the precipitous drop in employment owing to increasing efficiency; and the consequently inevitable collapse of the "Price System" (conventional free market economics), a term popularized by the Technical Alliance's theorist in Veblen, *Engineers and the Price System*. Graphs are prominent in successive editions of the Technocracy Study Course and in vue-graph transparencies and exhibition placards employed between the 1930s and 1990s. For collections of exhibition and lecture materials, see UBCA RBSC-ARC-1549, box 2, UAA 96-124-5 to 8, and UAA 96-127-8.

39 Scott, *Words and Wisdom of Howard Scott*.

40 Hubbert was active in technocracy while studying geology, physics, and mathematics at the University of Chicago, gaining a BSc (1926), an MA (1928), and a PhD (1937). While completing his doctorate he worked as a petroleum company geologist and taught a course on geophysics at Columbia University, first meeting Howard Scott in a Greenwich Village club in 1931. With others from the discussion group, they formed Technocracy Inc., and, as Hubbert recalled, he "drew up a kind of a small study course of the basics of what we were talking about, for use in these small groups that were assembling around." As press coverage of technocracy waned, Hubbert concluded that the organization "wasn't going to accomplish anything I was interested in. The technical part of it simply wasn't going anywhere," and so he "put it on the shelf." See Hubbert and Doel, "Oral History," but compare these reminiscences with note 53 below. From 1943 until 1964, Hubbert was a geologist for Shell Oil Company. He was later a professor of geology at Stanford University and then at UC Berkeley, and a research geophysicist for the United States Geological Survey from 1973 to 1976. He is best known for the "Hubbert curve," his 1956 prediction of "peak oil," according to which oil production will reach a maximum followed by a precipitous decline.

41 Hubbert, "Lesson 22," 242. The example is reminiscent of the discussion in Winner, "Do Artifacts Have Politics?"

42 Huxley, *Brave New World*.

43 Hubbert, *Technocracy Study Course* (1934), with successive editions in 1937, 1938, and 1940, and multiple reprintings and abridged editions after 1970.

44 The original source and exemplars of the illustration appear to be unknown

by current administrators but probably date from the mid-1930s, when
Technocracy Inc. generated publicity materials. The top two vehicles are
consistent with electric trolley car designs in operation between the 1890s
and 1910s. The bottom-most form could be as early as the Peter Witt trolley
of 1916 but is reminiscent of the 1936 PCC design. Streetcars were increasingly replaced by buses from the 1930s, however, and disappeared from most
North American cities by 1960.

45 George Wright to author, e-mail, 26 February 2016. The iconic illustrations
generated by the "Energy Survey" of the Technical Alliance and subsequent
research by Technocracy Inc. appear repeatedly in presentations to local audiences, public exhibits, and the higher-budget regional publications (e.g.,
Technocracy, *Technocrat*, *Northwest Technocrat*, and *Technocracy Digest*, each
published from the mid-1930s). Fewer illustrations accompanied the typically typewritten and mimeographed newsletters that included, at best,
hand-sketched line drawings. Among the most common illustrations were
graphs of rising production, energy consumption, and technology capacity
such as railway miles; a hierarchical organization chart of the planned
"North American Technate"; a map of the area of its intended coverage, consisting of Greenland, Canada, the United States, the Caribbean, and Central
and northern South America; and, the "Energy Certificate," an IBM-like
tabulator card intended to replace money with an accounting of energy
allocation. See UAA Technocracy Fonds.

46 Urquart, "Ring out the Old."

47 Bounds, "What's Yours Is Mine," 7.

48 Palm, "Why North America Faces Social Change," 9.

49 L.L.B., "Subsidies and Sabotage," 22.

50 The use of outmoded graphics was not likely intended as nostalgic appeals
to long-standing members but, instead, reflected a lack of contemporary
research by the organization and the image's satisfactory effectiveness in
attracting fresh audiences.

51 Dickinson, *Technocracy Digest*, 1.

52 CHQ Technocracy Inc., "Symbolization of Technocracy"; Smith, "'Symbolization Drives."

53 Hubbert's ineligibility followed the Civil Service Commission's concerns
about Technocracy Inc. The chair enquired pointedly about its reputed
racial exclusivity, internal governance, and rhetoric of intelligent engineers
to replace elected officials. As Hubbert reported privately to Howard Scott,

he hoped to appeal the decision through presumably like-minded contacts, whom he identified as former assistant attorneys general, an attorney for the Railroad Brotherhoods, and a counsel of the National Recovery Administration. However, the commission confirmed that he was unsuitable for federal employment because "Technocracy Inc is Fascistic in its setup and objectives and envisages a form of government not democratic in its character." See Civil Service Commission, "Questioning of M. King Hubbert, Division of Supply and Resources, before the Board of Economic Warfare," n.p.; Hubbert to Howard Scott, letters, 3 April, 6 May, 16 July, and 23 August 1943; US Civil Service Commission to Hubbert, letter, 14 August 1943, all from Hubbert Papers, Technocracy Continental Headquarters, Ferndale WA, reproduced in http://www.hubbertpeak.com/hubbert/Technocracy1943.pdf (viewed 4 April 2019 but no longer extant). I thank Mason Inman, author of *Oracle of Oil*, for confirming provenance.

54 Adair, "Technocrats 1919–1967." A film of the event made for internal consumption documents its meticulous organization and "long line of grey cars … extending nearly ten miles back from the Canadian border." See Technocracy Inc, "Operation Columbia," https://www.youtube.com/watch?v=RU regwsi_cw.

55 Adair, "Technocrats," 101.

56 For example, "Technocracy Demonstration in San Francisco"; "Operation Bakersfield"; Technocracy Inc., "Operation Ohio Valley."

CHAPTER THREE

1 Elsner Jr, "Messianic Scientism," 65.

2 Scott, "Origins of Technical Alliance." The congruence of technocratic thinking and management of early nuclear energy was noted by contemporaries, for example, "Atom Seen Causing 'New Technocracy.'"

3 Kennedy, *Engineers of Victory*; Edgerton, *Britain's War Machine*; King and Kutty, *Impact*.

4 Sheldrake, "Henry Gantt."

5 Hodges, *Alan Turing*; Smith, *Station X*.

6 Kirby, *Operational Research in War and Peace*; Waddington, *Operational Research in World War II*.

7 Rhodes, *Making of the Atomic Bomb*.

8 Kevles, *Physicists*; Galison and Hevly, *Big Science*. For a popular account, see Burchard and Killian, *Q.E.D.*

9   Johnston, "Security and the Shaping of Identity."

10 For example, Anderson, *Dam Busters*. See also Frayling, *Mad, Bad and Dangerous*. The revelation decades later of the wartime code-breaking work at Bletchley Park revived the trope of the introvert technologist in films such as Wise, *Breaking the Code*; and Tyldum, *The Imitation Game*.

11 McDougall, *Heavens and the Earth*; Coleman, "War on Cancer."

12 Del Sesto, "Wasn't the Future of Nuclear Energy Wonderful?"; Frenkel, "Hot Idea?"; Buchanan, "Atomic Meal."

13 Contemporary popular accounts include Böttcher, *Wonder Drugs*; Rowe, *One Story of Radar*.

14 On their short-lived political and policy aspirations, see Smith, *Peril and a Hope*; Strickland, *Scientists in Politics*. For first-hand perspectives see, for example, Skinner, "Atomic Energy and the Public Interest"; and Mott, "Scientist and Dangerous Thoughts."

15 Bush, *Science*, pt. 3, p. 18.

16 Wiener, *Cybernetics*.

17 For example, an intuitive notion of technological approaches to mob control was proposed by Alvin Weinberg regarding the Los Angeles riots, discussed in chapter 4.

18 For a rare collection of essays on the topic, see Hein, Diefendorf, and Ishida, *Rebuilding Urban Japan*.

19 On the rebuilding of Warsaw, Dresden, and French cities, see Diefendorf, *Rebuilding Europe's Bombed Cities*.

20 London County Council, *Proud City*.

21 Bruce, "First Planning Report"; Abercrombie and Matthew, "Clyde Valley Regional Plan."

22 Lutyens and Abercrombie, "Plan for the City"; Jones, "Fairer and Nobler City."

23 Abercrombie, "Civic Society," 80.

24 Weinberg and Stow, "Interview."

25 Lanouette and Silard, *Genius in the Shadows*, 400. Szilard, Meier, and Weinberg all mused about technologies to reduce human fertility. See, for example, Meier, *Modern Science and the Human Fertility Problem*. Meier and Weinberg each reviewed Szilard's career publications. See Feld and Szilard, *Collected Works of Leo Szilard*.

26 For example, Meier, *Planning for an Urban World*.

27 Meier, *Science and Economic Development*, 226.

28 Ibid., v.

29 Ibid., 1.

30 Ibid., vi and viii.

31 Ibid., 2–3.

32 Ibid., 36–7. On more ambitious enriched food technologies, see chapter 8 (this volume).

33 The MPF was developed by biochemist Henry Borsook (1897–1984) at the request of Clifford Clinton, a restaurateur, who distributed the low-cost food via his Meals for Millions international development organization between 1946 and 1956.

34 Meier, *Science and Economic Development*, 44.

35 Ibid., 100–1.

36 Ibid., 110.

37 Ibid., 112.

38 Meier, "World-Wide Prospect," 143.

39 Meier, *Science and Economic Development*, 135.

40 Ibid., 140.

41 Kirkwood and Price, "Technology-Enhanced Learning"; Sims, *Disruptive Fixation*.

42 For example, Warschauer and Ames, "Can One Laptop Per Child Save the World's Poor?"

43 Meier, *Science and Economic Development*, 141. Fleischer's film *Soylent Green* (1973) was based on Harry Harrison's novel *Make Room! Make Room!* The world population according to Meier's 1956 estimate was 2.7 billion, about one-twentieth of his estimated maximum carrying capacity. Meier's own estimate for the year 2020 was a population of 3 to 5 billion, considerably less than the actual value of some 7.8 billion. The growth rate has declined significantly since the late 1960s but is currently projected to reach about 11 billion in 2100. See http://www.worldometers.info/world-population/.

44 Meier, *Modern Science and the Human Fertility Problem*. See also Marks, *Sexual Chemistry*.

45 Meier, *Science and Economic Development*, 194 (emphasis added).

46 Ellul, *La Technique*, translated a decade later as *The Technological Society*.

47 Tugwell, "One World – One Wealth," 194. Tugwell (1891–1979) was closely associated with the Roosevelt administration's "New Deal" public programs of the 1930s and, later, was a policy and planning academic.

48 Levitt, "Review of Meier," 171.

49 Cottrell, "Review of Meier," 261, 262, 263.

50 Cairncross, "Freeing the World from Want," 36. On Wells, see chapter 5 (this volume).

51 Baran, "Review of Meier."

52 Velardo, "Review." Another summarized Meier's "ponderous and unexciting style." See Tomasson, "Review of Meier," 278.

CHAPTER FOUR

1 Eisenhower, "Farewell Address." See also Leslie, *Cold War and American Science*; Galison and Bernstein, "Physics between War and Peace."

2 The interdisciplinary field of "Science and Technology Studies," or "Science, Technology, and Society" (STS), combines history, philosophy, sociology, and political theory to investigate the interactions between science and society – that is, the perspective adopted in this book.

3 Supervised by Rashevsky, Weinberg decided that theorizing was premature and sought to do experimental postdoctoral studies in biophysics, but he was diverted towards nuclear physics just before the United States entered the Second World War. See Weinberg, *First Nuclear Era*, 7–9.

4 Wigner and Szanton, *Recollections of Eugene P. Wigner*, 216–17; Weinberg, *First Nuclear Era*, 13.

5 Oak Ridge Operations Manager, "Dr Alvin A. Weinberg Security Clearance Meeting 29 Sep 1948," Oak Ridge National Laboratory, MPA.0332, box 14, fol. 4.

6 For example, Alvin M. Weinberg to J.R. Oppenheimer, 6 January 1948, MPA.0332, box 17, fol. 5.

7 Hewlett and Duncan, *Nuclear Navy*.

8 On the growth of the occupation, profession, and academic discipline of nuclear engineering, see Johnston, *Neutron's Children*.

9 Alex Zucker to author, interview, 16 May 2016.

10 See, for example, Wigner and Szanton, *Recollections of Eugene P. Wigner*, 287–96; Weinberg, *Reflections on Big Science*, 139–40.

11 Alvin M. Weinberg to H.L. Brode, letter, 1 April 1965, CMOR, cab. 2, drawer 4, Civilian Defense File.

12 CMOR, cab 14, drawer 4, Harry D. Smyth File.

13 Branscomb, "Harvey Brooks," 463. See, for instance, Brooks, "Evolution of US Science Policy."

14 Alvine M. Weinberg to Jerome Weisner, memo, 26 June 1961, CMOR, cab. 3
    drawer 4, De Solla Price Correspondence. Weisner (1915–1994) was an MIT
    professor of engineering and chair of President Kennedy's Science Advisory
    Committee.

15 For example, "Cure Cancer? Go to Mars?"; Weinberg, "Big Science: A Liabil-
    ity?"; "Big Science – Marvel or Menace?"; Bylinsky, "Is 'Big Science' Headed
    for 'Big Trouble'?" The speech (Weinberg, "Impact of Large-Scale Science on
    the USA," MPA.0332, box 112, fol. 2, paper 29) was described as "frank and
    courageous" by his local newspaper and, like many of his subsequent public
    addresses, was published as an essay. See Weinberg, "Impact of Large-Scale
    Science."

16 De Solla Price, "Acceleration of Science"; "Beginning and the End of the Sci-
    entific Revolution"; Alvin M. Weinberg to Derek De Solla Price, 29 May 1963,
    CMOR, cab. 3, drawer 3, De Solla Price Correspondence; De Solla Price, *Little
    Science, Big Science*; Weinberg, *Reflections on Big Science*; Alvin M. Weinberg
    to Lord Snow of Leicester, 24 August 1965, CMOR, cab 5, drawer 4, Chron
    1965-2.

17 Mesthene, "On Understanding Change."

18 Cukier and Mayer-Schönberger, "Dictatorship of Data"; Cohen-Cole, *Open
    Mind*.

19 Finkbeiner, *Jasons*, 72. Wang, *In Sputnik's Shadow*. Department of Defense
    research projects such as Project MICHIGAN (1954–69) invested heavily in
    research exploring potential technologies of battlefield surveillance to iden-
    tify enemy incursions but little on social aspects of warfare. See Johnston,
    Holographic Visions.

20 These organizations were recent creations. PSAC had been formed by Eisen-
    hower in 1957 from the earlier Science Advisory Committee (SAC) of the
    Office of Defense Mobilization (ODM), 1951. JASON appeared in 1959. The
    Cambridge Discussion Group began in 1966, described by member John
    Kenneth Galbraith as a group "derived from the original Scientists and Engi-
    neers for Johnson ... who have been meeting on the problems of Viet Nam,"
    having organized before the 1964 elections to oppose conservative Republi-
    can Barry Goldwater. See Bridger, *Scientists at War*, 116. See also Greenberg,
    Science, Money and Politics; Bromley, *President's Scientists*; Weinberger,
    Imaginary Weapons.

21 A more recent counterpart in American government was Defense Secretary

Donald Rumsfeld, who argued that precision weapons could largely replace soldiers and re-establish democracy in the second Gulf War (2003). See, for example, Boot, "New American Way of War"; Rasmussen, *Risk Society at War.*

22 Weinberg, "Social Problems and National Socio-Technical Institutes."

23 Weinberg drew on contemporary critiques to nuance his own messages. These updated the warnings of John von Neumann about large-scale technologies, cited the work of biologist Rachel Carson on the ecological catastrophe of DDT, and chimed with ideas expressed by Hans Bethe scarcely a month earlier. See von Neumann, "Can We Survive Technology?"; Carson, *Silent Spring*; Bethe, "Social Responsibilities of Scientists and Engineers"; Weinberg, "Technological Gigantism."

24 Weinberg recommended the catch-phrase "Water for Peace" to describe nuclear desalination research, modelled on the earlier phrase "Atoms for Peace" coined by the Eisenhower administration, "a phrase that has been widely used and which has brought prestige to our country even if it hasn't brought peace to the world." See Alvin M. Weinberg to David Z. Beckler, 22 September 1965, CMOR, cab. 5, drawer 4, Chron 1965-2. Press praise includes, "The University and Society," *New York Times*, 10 August 1965.

25 Alvin M. Weinberg to Eugene Garfield, 26 January 1967, CMOR, cab. 5, drawer 4, Chron 1967-1.

26 Alvin M. Weinberg to E.W. Morehouse, 15 March 1966, CMOR, cab 5, drawer 4, Chron 1966-1; Alvin M. Weinberg to Garrett Hardin, 25 March 1966, CMOR, cab 5, drawer 4, Chron 1966-1; Hardin, "Lifeboat Ethics."

27 Newman, "Calls Dread of the Atom Bar to War."

28 For a contemporary account for the *Los Angeles Times*, see Cohen and Murphy, *Burn Baby Burn!*

29 Alvin M. Weinberg to Donald F. Hornig, 19 August 1965, CMOR, cab. 5, drawer 4, Chron 1965-2.

30 Alvin M. Weinberg to John A. McCone, 23 August 1965, CMOR, cab 5, drawer 4, Chron 1965-2; Weinberg and James C. Bresee, "Intended for 17th Conf on Sci and World Affairs, Ronneby, Sweden 3–8 Sep 1967," CMOR, cab 5, drawer 4, Chron 1968-1.

31 The "someone" providing the reference (see Huntington, *Civilization and Climate*) was likely Harvey Brooks. Weinberg recommended him to Mesthene as "a superb person to advise you about the validity of the idea" and

subsequently highlighted Brooks's role in shaping it during their conversations between Boston meetings. See Alvin M. Weinberg to Joshua Lederberg, 24 March 1969, CMOR, cab. 5, drawer 4, Chron 1969; Weinberg, *First Nuclear Era*, 150.

32 On social power and reproduction, see Rose and Hanmer, "Women's Liberation"; and Stabile, *Feminism and the Technological Fix.*

33 Alvin M. Weinberg to Emmanuel G. Mesthene, 19 August 1965, CMOR, cab. 5 drawer 4, Chron 1965-2.

34 Alvin M. Weinberg to Emmanuel G. Mesthene, 16 November 1965, CMOR, cab. 5 drawer 4, Chron 1965-2; Alvin M. Weinberg to Emmanuel G. Mesthene, 1 February 1966, CMOR, cab. 5, drawer 4, Chron 1966-1.

35 Nader, *Unsafe at Any Speed.*

36 Alvin M. Weinberg to Ralph Nader, 7 June 1966, CMOR, cab. 5, drawer 4, Chron 1966-2.

37 Between 1963 and 1966, Claire Nader directed Science in Society Studies at the Oak Ridge Institute of Nuclear Studies (ORINS). Weinberg's sister Fay (1910–2010) was also a sociologist; she worked at the University of the Pacific in Stockton, California.

38 Alvin M. Weinberg to Charlotte S. Read, 2 February 1966, CMOR, cab. 5, drawer 4, Chron 1966-1; Weinberg, "Will Technology Replace Social Engineering?"; "Can Technology Replace Social Engineering?," (emphasis added).

39 Reproduced in *Oak Ridge National Laboratory News* (1966), *Scientific Research* (1966), *University of Chicago Magazine* (1966), *Bulletin of the Atomic Scientists* (1966), *Air Force and Space Digest* (1967), *American Behavioral Scientist* (1967), and the *Chicago Jewish Forum* (1967). The unusually wide dissemination of the essay owed much to Weinberg's active efforts to publish the text quickly via receptive editors. See, for example, Alvin M. Weinberg to Gerald Gordon, 6 July 1966, CMOR, cab. 5, drawer 4, Chron 1966-2; Alvin M. Weinberg to Eugene Rabinovitch, 5 August 1966, CMOR, cab. 5, drawer 4, Chron 1966-2.

40 "Weinberg Touts 'Technological Fixes' to Stabilize World; Weinberg, "Can Technology Stabilize the World Order?"; "Is Technology the Answer to All Our Problems?"; Alvin M. Weinberg to James E. Spicer, 22 November 1966; Alvin M. Weinberg to Jaro Mayda, 9 December 1966; Alvin M. Weinberg to John D. Winebrenner, 23 December 1966. The latter three items in CMOR, cab. 5, drawer 4, Chron 1966-2.

41 Alvin M. Weinberg to Harvey Brooks, 17 June 1966, CMOR, cab. 5, drawer 4, Chron 1966-2.

42 Meier, *Science and Economic Development*, 139.

43 Alvin M. Weinberg to Richard L. Meier, 19 December 1966, CMOR, cab. 5, drawer 4, Chron 1966-2.

44 Ralph Nader to Alvin M. Weinberg, 22 October 1966, CMOR, cab 7, drawer 1, Nader.

45 For example, Wood, "Birth of the Technical Alliance"; Scott, "Public Lecture by Howard Scott"; "Design, Direction or Disaster."

46 Hubbert M. King to Alvin M. Weinberg, 19 September 1961, CMOR, cab. 6, drawer 1; Hubbert M. King to W.T. Thagard and Alvin M. Weinberg, 19 September 1961, CMOR, cab. 6, drawer 1; Hubbert M. King to Alvin M. Weinberg, 31 March 1967, CMOR, cab. 5, drawer 4, Chron 1967-1, mentioning face-to-face encounters and topics of energy production and national resources. Both Hubbert and Weinberg were members of the Washington, DC, Cosmos Club.

47 Oak Ridge Operations Manager, "Dr Alvin A. Weinberg Security Clearance" MPA.0332 box 14, fol. 4.

48 Weinberg, "Beyond the Technological Fix," 1; *First Nuclear Era*, 150 (emphasis in original).

49 Alvin M. Weinberg to Harvey Brooks, 17 June 1966, CMOR, cab. 5, drawer 4, Chron 1966-2.

50 For a lucid exposition of Weinberg's views, see his collected essays and auto-biography, Weinberg, *Nuclear Reactions* and *First Nuclear Era*.

51 Scott, "History and Purpose of Technocracy," 9.

52 Ibid., 11.

53 "Social engineering," for Weinberg, was a catch-all and rather derisive term that included sociologically informed interventions, legislation aimed at controlling behaviours, and even religious instruction. By contrast, the term connotes conventional engineers solving social problems in Jordan, *Machine-Age Ideology*.

54 Weinberg, "Can Technology Replace Social Engineering?," 5. A more extended version of the claim was made by economist and national security adviser to President Johnson Walt Rostow (1916–2003), in Rostow, *Stages of Economic Growth*.

55 Scott, "Public Address by Howard Scott."

56 Weinberg, "Can Technology Replace Social Engineering?," 7.

57 Ibid., 5.

58 Veblen, "Memorandum on a Practical Soviet of Technicians," 86–104.

59 Wood, "Birth of the Technical Alliance."

60 Ivie, *America Must Show the Way!*

61 Weinberg, "Can Technology Replace Social Engineering?," 5 and 8. See also
    Weinberg, "Social Problems and National Socio-Technical Institutes";
    "Beyond the Technological Fix."

62 In this respect, the communications of Scott and Weinberg were similar
    to those of F.W. Taylor, whose case for Scientific Management was founded
    on anecdotally recounted case studies.

63 Weinberg, "Basic Research and National Goals"; "Department of Science."

64 Barber, "Review of *Reflections on Big Science*," 170.

65 Alvin M. Weiberg to Harvey Brooks, 3 March 1967, CMOR, cab. 5, drawer 4,
    Chron 1967-1; Weinberg, "Social Problems and National Socio-Technical
    Institutes," 415.

66 Ibid., 417.

67 Ibid., 416 (emphasis in original).

68 Ibid.; Alvin M. Weinberg to William H. Davenport, 14 June 1967, CMOR, cab.
    5, drawer 4, Chron 1967-1; Kimbrell, "Weinberg-Baker Propose Socio-Techno
    Institutes."

69 Weinberg and Bresee," On the Air-Conditioning of Low-Cost Housing,"
    CMOR, cab. 5, drawer 4, Chron 1968-1, 16 January 1968; "NCHA Plans Public
    Housing Air-Conditioning," *Washington Post*, 5 January 1968; Alvin M.
    Weinberg to John S. Foster Jr, 7 March 1967; Alvin M. Weinberg to William
    P. Steven, 29 April 1967; "Agenda for the Seventies – Commencement Address
    for Alfred University (Revised, N.D.)." Latter three in CMOR, cab. 5, drawer 4,
    Chron 1967-1. Weinberg's concept had, in fact, already been thoroughly in-
    vestigated as the "McNamara Line" described above, in part by colleagues
    whom he knew well at MIT and as a member of PSAC.

70 "Intended for 17th Conference on Science and World Affairs, Ronneby,
    Sweden 3–8 September 1967," CMOR, cab. 5, drawer 4, Chron 1967-2.

71 Small nuclear desalting facilities had been operated by the Soviet Union
    from 1964, and later were built by Japan and India.

72 Weinberg, "Technological Fixes, Carbon Dioxide and Water," MPA.0332, box
    114, fol. 24; Eisenhower and Strauss, "Proposal for Our Time."

73 Alvin M. Weinberg to Lewis J. Strauss, telegram, 6 September 1968; "Memo-randum on Some Technological Possibilities for the Seventies 'Sent ... to Each of the Major Presidential Candidates.'" Both in CMOR, cab. 5, drawer 4, Chron 1968-2.

74 Mesthene, *Technology and Social Change*; Seaborg, "Toward a Science of 'Techumology.'" Weinberg published some twenty-three articles, reviews, and editorials in *Minerva* 1962–96; in 1983 he and Eugene Wigner were in-strumental in arranging funding that allowed the journal to continue publi-cation. On Weinberg's retrospective views, see Weinberg, "Edward Shils and the 'Governmentalisation' of Science."

75 Alvin M. Weinberg to Waldo E. Smith, 31 January 1967, CMOR, cab. 5, drawer 4, Chron 1967-1 (emphasis in original).

76 Barber, "Review of *Reflections on Big Science*."

77 Meier, "Social Impact of a Nuplex," 17.

78 Burns and Studer, "Reflections on Alvin M. Weinberg," 34–5 (emphasis in original).

79 Weinberg, "Response to Burns and Studer," 197, 199; Burns and Studer, "Reply to Alvin M. Weinberg."

80 Weinberg, "Technological Fixes and Social Fixes"; Club of Rome, *Limits to Growth*. Earlier examples broaching the topic include Weinberg, "Technol-ogy, Youth and the Environment"; "Towards a Compassionate Technology"; "Compassionate Technology."

81 "The Social Scientists," *Weekly People*, 12 November 1966. The review cited Weinberg's earlier position in Weinberg, "Effects of Scale on Modern Science and Technology."

82 See Baran, "Review of Meier," criticizing Meier's analysis as "reflecting naïve rationalism or the spirit of technocratic speculation" (1021); Akin, *Technoc-racy and the American Dream*; Burris, *Technocracy at Work*.

83 Weinberg, "Social Problems and National Socio-Technical Institutes," 433.

84 Oelschlaeger, "Myth of the Technological Fix."

85 Naess, "Shallow and the Deep"; Drengson, "Sacred and the Limits of the Technological Fix." On moral values and religious underpinnings of these wider critical perspectives, see Drengson, *Practice of Technology*.

86 Weinberg, "How Appropriate Is Appropriate Technology?"; "A Wake-up Call for Technological Somnambulists"; Winner, "Do Artifacts Have Politics?"

87 Weinberg, *First Nuclear Era*, 117, 183; "Technological Fixes, Carbon Dioxide

and Water," typescript, MPA.0332, box 114, fol. 24; Moriarty and Honnery, "Nuclear Energy."

88   Emanuel Epstein to Alvin M. Weinberg, 5 May 1971, CMOR, cab. 12, drawer 1.

89   The Institute of Energy Analysis was a unit of the Oak Ridge Associated Universities (ORAU) between 1976 and Weinberg's retirement in 1984.

90   Weinberg, "Social Institutions and Nuclear Energy"; "Salvaging the Nuclear Age"; "Inherently Safe Technologies (Chemical and Nuclear)."

91   Weinberg, "Sanctification of Hiroshima"; "Limits of Science and Trans-Science."

92   Alvin M. Weinberg to Edward Shils, 30 August 1988, MPA.0332, box 78, fol. 9; Weinberg, "Nuclear Power and Public Perception," 279.

93   Anthologies include Mesthene, *Technology and Social Change*; De Nevers, *Technology and Society*; Teich, *Technology and Man's Future*; Cross, Elliott, and Roy, *Man-Made Futures*; Thompson, *Controlling Technology*; Weinberg, *Nuclear Reactions*; Howard, *Environmental Policy Reader*; MacKenzie, *Science and Technology Today*; and Hawisher and Selfe, *Literacy, Technology, and Society*.

94   An important echo of Weinberg and Meier in government-supported private industry was Simon Ramo (1913–2016) of TRW Inc. (see Ramo, *Cure for Chaos*). See Dyer, "Limits of Technology Transfer."

95   Meier, "Social Impact of a Nuplex," 16; Burns and Studer, "Reflections."

96   For instance: "most technological fixes can do no more than help remedy the immediate problem that invoked the fix. In their wake they leave other problems that, in turn, are amenable to resolution by additional technological fixes: fixes are applied over fixes, and the society, to be metaphorical, becomes a patchwork of band-aids – indeed, I have referred to it as the "band-aid society." See Weinberg, "Beyond the Technological Fix," 3.

97   Weinberg, *First Nuclear Era*, 151.

98   For example, this redefinition of technocracy infuses Toffler, *Future Shock*, 404–40. John Taube, the principal publicist of Technocracy Inc. in the post-Scott period, actively sought to correct the conflation of the characteristics of his organization with other expressions of technologist-dominated governing elites such as the Kremlin. See, for example, Taube to Seweryn Bialer and Joan Afferica, 4 August 1986, UAA 96-123-5, fol. 149.

99   The course materials long served as the standard instruction for members of the organization, being revised five years after Howard Scott's death as

Technocracy: Technological <u>Social</u> Design (Savannah, GA: Continental Headquarters, 1975), revised and retitled *Technocracy: Technological <u>Continental</u>
Design* (Ferndale, WA: Continental Headquarters, 2001) (emphasis added),
and reformatted for internet distribution in 2004. The streetcar example
was incorporated for the first time as a graphic in the abridged post-1970
versions.

100 John A. Taube to Isaac Asimov, 6 February 1980, UAA 96-123-5, fol. 143;
     John A. Taube to Ben Bova, 18 August 1982, UAA 96-123-5, fol. 146; Hardin,
     "Tragedy of the Commons"; "Lifeboat Ethics."

101 Garrett Hardin to John A. Taube, 18 October 1981, UAA 96-123-5, fol. 144;
     Robert Schaeffer to John A. Taube, 11 October 1983, UAA 96-123-5, fol. 146.
     Media coverage increasingly portrayed technocracy as an outmoded faith
     preserved by old men. See Livingston, "Technocracy Still Lives"; Maloney,
     "Technocracy Dreams at the Fringe"; Hawthorn, "Diehard Few Keep
     Utopian Dream Alive."

102 Walt Fryers to John A. Taube, 22 February 1988, UAA 96-123-4, item 135.

CHAPTER FIVE

1 Bellamy, *Looking Backward.*

2 Scientific socialism was described by Friedrich Engels as a testable scientific
   hypothesis, unlike what Marx had criticized as earlier forms of "utopian socialism" having no clear understanding of the contexts that produced social
   change. See Engels, *Socialism.*

3 Bellamy, *Equality.*

4 Alkon, *Science Fiction before 1900.*

5 Wells, *Time Machine.*

6 Wells, *Outline of History; World Set Free.* See also Bowler, *History of
   the Future.*

7 Wells, *Sleeper Wakes; Modern Utopia; Men Like Gods.*

8 Morris, *News from Nowhere;* "Bellamy's Looking Backward."

9 Wells, *Outline of History; Work, Wealth and Happiness of Mankind.*

10 Wells, *Shape of Things to Come.*

11 Menzies, "Things to Come."

12 Aldous Huxley to Kethevan Roberts, 18 May, cited in Huxley, *Letters of
    Aldous Huxley,* 348; *Brave New World.*

13 An excellent overview is Bowler, *Science for All.*

14 Dizer, *Tom Swift and Company*; Johnson, *Edward Stratemeyer and the Strate-meyer Syndicate*; Molson, "Boy Inventor in American Series Fiction."

15 Johnston, "Vaunting the Independent Amateur."

16 Massie and Perry, "Hugo Gernsback and Radio Magazines."

17 Gernsback, "Wonders of Technocracy," cited in Ashley, *Gernsback Days*, 206. See also Westfahl, *Hugo Gernsback and the Century of Science Fiction*.

18 Bradbury, "Editorial" (Summer 1939). See also Armytage, *Yesterday's Tomor-rows*, 132.

19 Yerke, "Revolt of the Scientists"; Reynolds, "Don't Get Technatal." On Reynolds as pseudonym, see Nolan, *Ray Bradbury Companion*, 299–300.

20 Bradbury, "Editorial" (Fall 1939); Weller, *Listen to the Echoes*.

21 Knight, *Futurians*, 47. See also Asimov, *In Memory Yyet Green*.

22 Cashbaugh, "Paradoxical, Discrepant, and Mutant Marxism." The need to reshape science fiction as a political medium was communicated in a speech and subsequent pamphlet. See Michel, "Mutation or Death!" See also Winter, "Michelist Revolution."

23 Michel, "What Is Science Fiction Doing for You?"; Cashbaugh, "Paradoxical, Discrepant, and Mutant Marxism," 68.

24 This broadly positive portrayal of technology in science fiction was suc-ceeded by the dystopian visions of the cyberpunk genre, characterized by the writing of William Gibson.

25 Asimov, *Foundation*; *Foundation and Empire*; and *Second Foundation*.

26 Future-fiction writers nevertheless began to factor Cold War pessimism into their works. George Orwell (1903–1950) described governance via technolo-gies that cowed the populace and that rewrote history; Kurt Vonnegut (1922–2007) depicted a more benign but insidious technocratic and automated civilization in which engineers and managers ruled a directionless public. Both extrapolated plausible futures from their own experiences: Orwell as wartime BBC staffer and literary journalist, and Vonnegut as postwar publi-cist for General Electric. See Orwell, *1984*; Vonnegut, *Player Piano*.

27 Pohl, *Way the Future Was*. See also Terzian, "The 1939–1940 New York World's Fair."

28 Kargon et al., *World's Fairs on the Eve of War*.

29 Norman Bel Geddes quotation in unpublished materials, cited in Morshed, "Aesthetics of Ascension," 77; Marchand, "Designers Go to the Fair 2." In New York State itself, the expressway system was implemented by

Commissioner Robert Moses (1888–1981), who selected the site for the fair. His role in shaping public access to New York's parks by filtering out low-income populations is described in Winner, "Do Artifacts Have Politics?"

30 Chandler, "First Industrial Exposition."

31 Auerbach, *Great Exhibition of 1851*; Greenhalgh, *Ephemeral Vistas*; Harriss, *Eiffel Tower*. See also Rieger, "Envisioning the Future."

32 Rydell, *World of Fairs*; Silla, "Chicago World's Fair of 1893."

33 Ganz, *1933 Chicago World's Fair*, 3; Schrenk, *Building a Century of Progress*.

34 Krige, "Atoms for Peace"; Johnston, *Neutron's Children*, chap. 4.

35 Pluvinge, *Expo 58*.

36 Smith, *Eisenhower in War and Peace*.

37 Rydell, Findling, and Pelle, *Fair America*.

38 Cotter, *Seattle's 1962 World's Fair*.

39 Kelly, *Expanding the American Dream*. A contemporary song disparaging such urban conformity was "Little Boxes," written by Malvina Reynolds (1962) and recorded by Pete Seeger the following year.

40 Hanna and Barbera, *Jetsons*; Taurog, *It Happened at the World's Fair*. A contemporary British counterpart was *Thunderbirds*, a science-fiction adventure series set a century in the future (ITV: ITC Entertainment, 1964–66). Featuring marionette characters of the International Rescue Organization, it centred on technological fixes: reliance on advanced spacecraft, aircraft, and ground transport vehicles to speedily resolve human crises. The creators of the series, Gerry and Sylvia Anderson, had led up to the series by producing children's programs with similar technological premises: *Supercar* (ITC Entertainment, 1961–62), *Fireball XL5* (ITC Entertainment, 1962–63), and *Stingray* (ITC Entertainment, 1964–65).

41 Samuel, *End of the Innocence*.

42 See Jacobsen, *Operation Paperclip*.

43 Haber, *Walt Disney Story of Our Friend the Atom*.

44 Foster, "Futuristic Medievalisms and the US Space Program."

45 Wiener, "There's a Great Big Beautiful Tomorrow."

46 Staudenmaier, "Recent Trends in the History of Technology."

47 Smith, "Back to the Future," 72–3.

48 Foster, "Futuristic Medievalisms and the US Space Program," 154. Foster labels Disney's vision as "technocratic populism," but there is little hint in these modern folktales of experts, hierarchies, or, indeed, politics.

49 Lownsbrough, *Best Place to Be*.

50 Keats, *You Belong to the Universe*; Turner, "Buckminster Fuller." Fuller wrote numerous books and autobiographical accounts. See, for example, Fuller and Marks, *Dymaxion World of Buckminster Fuller*. On a systems theory approach to life on earth, see his best known work, Fuller, *Operating Manual for Spaceship Earth*.

51 Greenhalgh, *Fair World*; Mattie, *World's Fairs*.

52 Wiener, "There's a Great Big Beautiful Tomorrow."

53 Cox, *American Radio Networks*; Barnouw, *Golden Web*.

54 Rhees, "New Voice for Science"; Terzian, *Science Education and Citizenship*.

55 Haring, *Ham Radio's Technical Culture*; Bureau of Standards, "Construction and operation of a very simple radio receiving equipment," Letter Circular LC 43, 16 March 1922, Smithsonian Institution Archives (SIA), RU7091, box 11, fol. 2; US Department of Agriculture and State Agricultural Colleges, "Cooperative extension work in agriculture and home economics – boys and girls radio clubs," n.d., c. 1922, SIA RU7091, box 11, fol. 3.

56 Nowlan, "Armageddon 2419 AD."

57 Aldiss and Wingrove, *Trillion Year Spree*.

58 Phillips, "Art for Industry's Sake."

59 An example of the style is figure 7.1, illustrating the hydroelectric projects of the Tennessee Valley Authority.

60 For example, Lucaites and Hariman, "Visual Rhetoric." This section samples and summarizes a longer discussion in Johnston, *Holograms*, chap. 3.

61 The latter advertisement is ironic, considering the company's role in the Bhopal chemical leak disaster, which, in 1984, exposed hundreds of thousands of inhabitants to toxic fumes in the shantytowns surrounding the Union Carbide plant.

62 Heumann, *Gegenstücke*; Benintende, "Who Was the Scientific American?"; Johnston, "Vaunting the Independent Amateur."

63 Herbert, *Mr Wizard's Science Secrets*; Lafollette, *Science on American Television*.

64 Spilhaus, "United States Science Exhibit"; Novak, "Sunday Funnies Blast Off into the Space Age."

65 Spilhaus, "Oral History Interview." On synthetic food, see Spilhaus, *Our New Age*, 14 November 1965. Samples of Radebaugh's imagery at https://www.flickr.com/search/?text=radebaugh.

66 Gabor, *Inventing the Future*; Brand, *Media Lab*.

67 Haber's career as an expositor of popular science in Germany is discussed in Heumann, *Gegenstücke.*

68 Wang, *American Science in an Age of Anxiety*; Wilson, "Conference address, Labour Party Annual Conference," Scarborough, 1963, which oriented British government policy towards harnessing the "white heat" of science and technology. See also Benn, "Government's Policy for Technology"; Powell, *Tony Benn.*

69 BBC4, *Mad and Bad: 60 Years of Science on TV.*

70 Baxter, Burke, and Latham, *Tomorrow's World*, 254, 257.

71 Boon, "Televising of Science."

72 Watt-Smith, *Big Life Fix with Simon Reeve*, season 1, episode 1.

73 Ibid., season 1, episode 3.

74 Trademarkia. "There's an App for That."

75 https://www.youtube.com/watch?v=khiyk26xTQ8.

76 Nextdoor Neighbourhood App (Nextdoor.com Inc.); EyeEm GmbH; Headspace (Headspace Meditation Ltd); Sleep Cycle Alarm Clock (Northcube AB). Hydration apps include Waterlogged (Day Logger Inc.), Aqualert (Premium Health and Fitness Apps), and Hydro Coach (Codium App Ideas), which incorporates eighty styles of drinking glasses, custom volumes, and drinks.

77 On the ethics of technological choice and usage, see Verbeek, *Moralizing Technology.*

78 There is an extensive and growing literature on library usage – for example, Vollmer, "There's an App for That!" See also Cummiskey, "There's an App for That," for health and physical education; Wellde and Miller, "There's an App for That!," on the transformative powers of social media for patient education and emotional support.

79 See, for example, Johnston, Franks, and Whitelaw, "Crowd-Sourced Science."

80 Buie and Blythe, "Spirituality."

81 Huntington, *Soldier and the State*, 457–8; Hodnett, *Art of Working with People*; Fox, *Engines of Culture*, 2. See also Hodnett, *Art of Problem Solving.*

82 Baker, *Library Media Program and the School*, 54.

83 "Volger Sees Grim Outlook."

84 FitzGerald, *Way out There in the Blue.* For a personal account by the chief scientist of the project, see Yonas and Gibson, *Death Rays and Delusions.*

85 American Physical Society Study Group et al., "Report."

CHAPTER SIX

1 Hughes, "Afterword."

2 Wells, "Wanted – Professors of Foresight!"

3 Stokes, "Wanted: Professors of Foresight in Environmental Law!"

4 Such technological nostalgia is typical of the Disney narrative discussed in chapter 5.

5 This perspective, referred to as "presentism" by historians, evaluates the past in terms of (unquestioned) current knowledge and cultural criteria. It is closely allied with "Whig historiography," the portrayal of past times as inferior and leading inexorably and progressively to the wisdom of the present age. See Johnston, *History of Science*, chap. 6.

6 Johnston, *Neutron's Children*, 261.

7 For a ready overview, see Bauman and Tillman, *Hitchhiker's Guide to* LCA.

8 For a historical case study of more common engineering experience that details side effects for human health, see Lecain, "When Everybody Wins Does the Environment Lose?"

9 Plans for the HS2 high-speed rail line from London to Birmingham, Manchester, and Leeds aiming to begin operation between the 2020s and 2030s, for example, have focused primarily on economics and potential environmental harms rather than on social benefits.

10 It is worth reflecting, for example, on your personal experience of the ongoing evolution of the internet, online consumption, and social media to assess how coherent your views are regarding their societal consequences.

11 Rosner, *Technological Fix*.

12 An excellent sociological study is Perrow, *Normal Accidents*. The investigation of specific accidents has also been a fertile source of sociotechnical studies, such as McConnell, *Challenger*; Schmid, *Producing Power*.

13 Johnson, Stokes, and Arndt, *Thalidomide Catastrophe*; Curran, "Thalidomide Tragedy in Germany." For a brief account by a key early investigator, see Lenz, "History of Thalidomide."

14 Most popular accounts of thalidomide have been written by survivors and families, rightly focusing on the neglected personal and social consequences. For a journalistic treatment, see *London Sunday Times* and Potter, *Suffer the Children*.

15 Perkins, "Reshaping Technology in Wartime"; Jarman and Ballschmiter, "From Coal to DDT"; Clarke, "Rethinking the Post-War Hegemony of DDT."

16 Bureau of Entomology and Plant Hygiene, "DDT," 1–2.

17 Schmitt, "From the Frontlines to Silent Spring"; Bartlett, "Chemical Marvels Take the Bugs Out of Living"; Whorton, *Before Silent Spring*.

18 Fry, "Studies of US Radium Dial Workers"; Moore, *Radium Girls*.

19 Michaels, "Waiting for the Body Count"; Sellers, "Discovering Environmental Cancer."

20 Bartrip, "History of Asbestos Related Disease."

21 Telford and Guthrie, "Transmission of the Toxicity of DDT"; Storer, "DDT and Wildlife"; Cottam and Higgins, "DDT"; Jarman and Ballschmiter, "From Coal to DDT."

22 Carson, *Sea around Us*. A film version (dir. Irwin Allen) won the 1953 Oscar for Best Documentary. Carson's first book was *Under the Sea Wind*, and the third of the trilogy was *The Edge of the Sea*.

23 Carson, *Silent Spring*.

24 The same evolutionary adjustment of ecosystems is responsible for the resistance of microorganisms to antibiotics.

25 Bookchin, *Our Synthetic Environment*.

26 On evaluations of Carson's contributions, see Lytle, *Gentle Subversive*; Murphy, *What a Book Can Do*.

27 Weinberg to William Hines, 27 July 1967, CMOR, cab. 5, drawer 4, Chron 1967-2.

28 The most optimistic estimates of filter tip efficacy were the earliest: Wunder, "Interrelationship of Smoking to Other Variables and Preventive Approaches"; Trichopoulos et al., "Lung Cancer and Passive Smoking." National statistics revealed a levelling off of cigarette consumption during the 1960s but a rate of lung cancer that continued to rise.

29 Novotny et al., "Cigarette Butts and the Case for an Environmental Policy."

30 Schuman, "Patterns of Smoking Behaviour"; Gendreau and Vitaro, "Unbearable Lightness of 'Light' Cigarettes."

31 Etzioni and Remp, *Technological Shortcuts to Social Change*, 3.

32 Larimer, "Treatment of Alcoholism with Antabuse."

33 Mom and Kirsch, "Technologies in Tension"; MacShane and Tarr, *Horse in the City*.

34 Walker, *Motorcycle*. A classic account of the social construction of the bicycle is Bijker, *Of Bicycles, Bakelites, and Bulbs*.

35 See, for example, Chamon, Mauro, and Okawa, "Mass Car Ownership in the Emerging Market Giants."

36 Setright, *Drive On!*; Kline and Pinch, "Users as Agents of Technological Change."

37 Wells, *Car Country*; Slater, "General Motors and the Demise of Streetcars."

38 Schrenk et al., "Air Pollution in Donora, Pa."

39 Haagen-Smit, "Chemistry and Physiology of Los Angeles Smog," 1,342; Jacobs and Kelly, *Smogtown*. The five-day Great London Smog of December 1952 caused an estimated four thousand excess deaths and 100,000 cases of respiratory illness, and led to the passing of the Clean Air Act in Britain to control coal burning. See Bell, Davis, and Fletcher, "Retrospective Assessment." Two similarly extreme episodes in eastern and northeastern China in 2013 punctuated rising pollution and government concern in Asian cities.

40 Kovarik, "Ethyl-Leaded Gasoline."

41 On air pollution more generally, see Uekoetter, "Solving Air Pollution Problems Once and for All."

42 Jacobson, *Atmospheric Air Pollution*. Lead credits were a policy predecessor of "carbon credits," which allowed companies, regions, or even countries to trade their legal allocation of lead pollution.

43 California Air Resources Board, "CARB History."

44 Kašpar, Fornasiero, and Hickey, "Automotive Catalytic Converters."

45 Høyer, "History of Alternative Fuels in Transportation."

46 Wakefield, *History of the Electric Automobile*; van Bree, Verbong, and Kramer, "Multi-Level Perspective."

47 For example, Brady, "Electric Car Cultures."

48 Franks, Hanscomb, and Johnston, *Environmental Ethics and Behavioural Change*, chap. 7.

49 Nye, "United States and Alternative Energies."

50 Casey, *Set Phasers on Stun*.

51 Google n-grams, https://books.google.com/ngrams/.

52 Wells, "Wanted – Professors of Foresight!"

53 Jakle and Sculle, *Lots of Parking*.

54 On technological momentum and lock-in, see Hughes, "Technological Momentum"; Foxon, "Technological Lock-in and the Role of Innovation."

55 Lutz and Fernandez, *Carjacked*.

56 Crowther et al., *Traffic in Towns*.

57 Mohl, "Stop the Road."

58 For an optimistic forecast, see McNichol, *Big Dig*; for a post-completion account, see Conti, "Boston Commute Is as Congested as It Was 10 Years Ago."

59 McNeish, "Resisting Colonisation."

60 Meikle, *American Plastic*.

61 "Plastics in 1940." See also Nichols and Leighton, "Plastics Come of Age."

62 For example, the conservative British tabloid the *Daily Mail* began a journal-istic campaign against plastic supermarket bags in 2008 and broadened it a decade later to include disposable plastic consumer items, coincident with documentaries by British nature broadcaster David Attenborough.

63 Rogers, *Gone Tomorrow*; Stokes, Köster, and Sambrook, *Business of Waste*.

64 Ryan, "Brief History of Marine Litter Research"; Sheavly and Register, "Marine Debris and Plastics"; Moore et al., "Comparison of Plastic and Plankton in the North Pacific Central Gyre."

65 See, for example, http://geo-dome.co.uk/article.asp?uname=problems; vom Saal and Hughes, "Extensive New Literature Concerning Low-Dose Effects."

66 Petkewich, "Technology Solutions."

67 European Commission, "European Strategy"; Oxo-Biodegradable Plastics Association, "OPA Responds."

68 Yousif and Haddad, "Photodegradation and Photostabilization of Polymers."

69 See Seaman, "Plastics by the Numbers."

70 For a psychologist's explanation for the holding of two contradictory views, see Kelman, "Compliance, Identification, and Internalization."

71 Hopewell, Dvorak, and Kosior, "Plastics Recycling."

72 For comparable effects in art and aesthetics, see Whiteley, "Toward a Throw-Away Culture."

73 Guffy, *Retro*.

74 Loose, "Maker Culture."

75 Etzioni and Remp, *Technological Shortcuts*, 1. See also Nelkin, *Methadone Maintenance*.

76 Tenner, *Why Things Bite Back*.

77 Alvin Weinberg introduced the phrase "Faustian bargain" to describe nu-clear energy and its long-lived dangers. See, for example, Weinberg, "Social Institutions and Nuclear Energy," 33.

78 Merton, "Unanticipated Consequences of Purposive Social Action."

CHAPTER SEVEN

1 Brown and Michael, "Sociology of Expectations," 1. See also Borup et al., "Sociology of Expectations in Science and Technology."

2 Opler, "Fourth-Generation Software."

3 Wirth, "Brief History of Software Engineering."
4 Backus et al., "Comparing Expectations to Actual Events." For a flavour of the hysteria, as well as accusations of inadequate interest from social scientists about technological side effects, see Peled, "Why Did Social Scientists Miss the Bug?"
5 Brooks, "No Silver Bullet." See also Ceruzzi, "The 'Problem' of Computer-Computer Communication." The terms "silver bullet" and "magic bullet" have exploded in popular usage since the 1980s.
6 See https://books.google.com/ngrams, "bug fix," "version number."
7 Weinberg, "Beyond the Technological Fix," 3.
8 Bug fixing has grown into an operational technology ironically reliant on human systems. See, for example, Bitzer and Schröder, "Bug-Fixing and Code-Writing."
9 Brooks, *Mythical Man-Month*.
10 Johnston, *Neutron's Children*, 50, 147, 218.
11 Egorov, *Radiation Legacy of the Soviet Nuclear Complex*; Ialenti, "Adjudicating Deep Time."
12 Cochran et al., "Fast Breeder Reactor Programs."
13 Nuclear and Radiation Studies Board, *Lessons Learned from the Fukushima Nuclear Accident*.
14 World Nuclear Association, "Storage and Disposal of Radioactive Waste," https://www.world-nuclear.org/information-library/nuclear-fuel-cycle/nuclear-waste/storage-and-disposal-of-radioactive-waste.aspx; Ojovan and Lee, *Introduction to Nuclear Waste Immobilisation*; Slovic, Flynn, and Layman, "Perceived Risk."
15 Pocock, *Nuclear Power*.
16 Weinberg, "Technological Fixes, Carbon Dioxide and Water"; "Social Institutions and Nuclear Energy"; *First Nuclear Era*, 150.
17 Johnston, *Neutron's Children*.
18 Cowan, *Social History of American Technology*; Gooday, *Domesticating Electricity*; Marvin, *When Old Technologies Were New*.
19 Simpson, *Dam!*; Oravec, "Conservationism vs. Preservationism."
20 Huxley, *TVA*; Staff Report, "TVA and the Fertilizer Industry"; Morgan, *Making of the TVA*; Colignon, *Power Plays*.
21 See, for example, the initial government promotion (Bourassa, *La Baie James*) and scholarly overviews (Bélanger and Comeau, *Hydro-Québec*).
22 For example, Sovacool and Bulan, "Behind an Ambitious Megaproject in

Asia"; Showers, "Congo River's Grand Inga Hydroelectricity Scheme";
Heggelund, *Environment and Resettlement Politics.*

23 This section draws upon portions of my chapter 4 in Franks, Hanscomb,
and Johnston, *Environmental Ethics and Behavioural Change.*

24 Leopold, *Sand County Almanac,* originally published in 1949; Flader, *Think-ing Like a Mountain.*

25 Leopold, "Land Ethic." A more recent example evincing similar sentiments is
McKibben, *End of Nature.*

26 Bookchin, "What Is Social Ecology?"

27 Ibid., "Social Ecology versus Deep Ecology," 9.

28 Naess, "Shallow and the Deep."

29 Naess and Rothenberg, *Ecology, Community, and Lifestyle;* Naess and
Drengson, *Selected Works of Arne Naess.*

30 Sujauddin et al., "Characterization of Ship Breaking Industry in
Bangladesh"; Ne er et al., "Shipbreaking Industry in Turkey.

31 Adger, "Social and Ecological Resilience; Cutter, "Landscape of Disaster
Resilience Indicators."

32 Weinberg to Waldo E. Smith, 31 January 1967, CMOR, cab. 5, drawer 4, Chron
1967-1.

33 For a narrowly technological approach to urban resilience, see Sitinjak et al.,
"Enhancing Urban Resilience through Technology and Social Media."

34 Weaver, "Can Technology Strengthen Our Resilience?"

35 American Red Cross and International Federation of Red Cross and Red
Crescent Societies, "Vision on the Humanitarian Use of Emerging Technol-ogy," 9.

36 A similar political perspective is at the heart of survivalist preparations, in
which personal survival through food stocks, weapons, and self-reliant facil-ities replaces confidence in community resilience and social traditions. See,
for example, Kabel and Chmidling, "Disaster Prepper."

CHAPTER EIGHT

1 Wozniak and Smith, *I, Woz.*

2 Apple Corp., "Advertisement."

3 For example, Pogačnik and Črnič, "Ireligion."

4 Kenney, *Understanding Silicon Valley.*

5 Anderson, *The Movement and the Sixties.*

6 Brand, *Whole Earth Catalog* (1968), extended through 1972 and then updated infrequently as *The Last Whole Earth Catalog*, *The Whole Earth Epilog* (1974), *The Next Whole Earth Catalog* (1980), *The Essential Whole Earth Catalog* (1986), *The Whole Earth Ecolog* (1990), *The Millennium Whole Earth Catalog* (1994), and so on.

7 See, for example, an account by one member: Rheingold, *Virtual Community*.

8 Brand, *Media Lab*. On the role of engineers in shaping human futures, see the much earlier example by Hungarian-British physicist-engineer Dennis Gabor, *Inventing the Future*.

9 Brand, *Whole Earth Catalog*, 1; Turner, *From Counterculture to Cyberculture*.

10 Day, Logsdon, and Latell, *Eye in the Sky*; Taubman, *Secret Empire*; Lewis, *Spy Capitalism*; McDougall, *Heavens and the Earth*.

11 For example, Rong et al., "Smart City Architecture"; Bompard et al., "Congestion-Management Schemes."

12 Jordan, "Haussmann and Haussmannisation"; Winner, "Do Artifacts Have Politics?"; Taylor, *Berlin Wall*.

13 Bradley, "Unbelievable Reality of the Impossible Hyperloop"; Strauss, "Elon Musk."

14 United Nations General Assembly, "Treaty on Principles."

15 Wells, *Men Like Gods*.

16 Leach, *Runaway World?*, 1, 16.

17 Brand, *Whole Earth Catalog*, 1.

18 Ibid., *Whole Earth Discipline*, 1, 21. For a more critical stance, see Preston, *Synthetic Age*.

19 Brand, *Whole Earth Discipline*, 208.

20 Ibid., 13, 21. For an authoritative history of anthropogenic climate change, see Weart, *Discovery of Global Warming*. See also Moriarty and Honnery, "Nuclear Energy."

21 Keith, "Geoengineering the Climate."

22 Schiermeier, "Iron Seeding Creates Fleeting Carbon Sink."

23 Carbon Capture and Storage Association, http://www.ccsassociation.org/.

24 Royal Society, "Geoengineering the Climate."

25 Wolpert, "No Quick, Easy Technological Fix for Climate Change."

26 National Institutes of Health, "Insulin and Human Growth Hormone"; Tomiuk, Wohrmann, and Sentker, *Transgenic Organisms*.

27 Redenbaugh et al., *Safety Assessment of Genetically Engineered Fruits and*

Vegetables; Bruening and Lyons, "Case of the Flavr Savr Tomato"; Tutelyan, *Genetically Modified Food Sources.*

28 Nicolia et al., "Overview of the Last 10 Years"; Committee on Genetically Engineered Crops, "Genetically Engineered Crops."

29 Brand, *Whole Earth Discipline*, 154–5; Potrykus, "Golden Rice and Beyond"; Mayer, "Golden Rice Controversy"; Dubock, "Politics of Golden Rice."

30 The growing literature on genetic engineering – like that on other forms of engineering – ranges from the narrowly technical to the more broadly questioning and interdisciplinary. See, for example, Setlow, *Genetic Engineering*; and Russo and Cove, *Genetic Engineering.*

31 Jordan and O'Riordan, "Precautionary Principle." On the problems with conventional twentieth-century technological systems, see Perrow, *Normal Accident.*

32 Brand, *Whole Earth Discipline*, 217.

33 Ibid., 217.

34 Huxley, "Transhumanism," 13.

35 Tirosh-Samuelson, "Technologizing Transcendence."

36 Fuller, *Humanity 2.0.*

37 For example, Dvorsky, "Better Living through Transhumanism"; Verdoux, "Transhumanism, Progress and the Future."

38 Spilhaus, "Our New Age," 26 December 1965.

39 Belasco, "Synthetic Arcadias."

40 Vinge, "Coming Technological Singularity"; Eden and Moor, *Singularity Hypotheses.*

41 Moore, "Cramming More Components onto Integrated Circuits"; Byrne, Oliner, and Sichel, "Is the Information Technology Revolution Over?"

42 Toffler, *Future Shock*; *The Third Wave*; Toffler and Toffler, *Creating a New Civilization.* See also Halley and Vatter, "Technology and the Future as History"; and De Miranda, "Technological Determinism and Ideology."

43 Berube, *Nano-Hype.*

44 Berger, *Nano-Society.*

45 Drexler, *Engines of Creation*; Smalley, "Of Chemistry, Love and Nanobots."

46 "Nanotechnology and the Environment"; Health and Environment Alliance, "Nanotechnology and Health Risks."

47 *Star Trek: The Next Generation* (1987–94) explores the moral worth of synthetic humans via the android character Data but rejects the transhumanist Borg as re-engineered but dehumanized beings.

48 Stern, *Eugenic Nation*.

49 Dikötter, "Race Culture"; Bashford and Levine, *Oxford Handbook of the History of Eugenics*.

50 Huxley, *Brave New World*.

51 Teles, "Kludgeocracy in America."

52 This section extends a discussion in Franks, Hanscomb, and Johnston, *Environmental Ethics and Behavioural Change*, chap. 4.

53 Steffen, *Worldchanging*.

54 "Shallow" and "deep" ecology can be distinguished at a philosophical level, with the first focusing on utilitarian ethics and the second founded on virtue ethics.

55 Schumacher, *Small Is Beautiful*, 128.

56 Kirk, "Machines of Loving Grace"; "Appropriating Technology."

57 https://books.google.com/ngrams.

58 Directorate-General for Research and Innovation, "Options for Strengthening Responsible Research and Innovation," 3; Owen, Macnaghten, and Stilgoe, "Responsible Research and Innovation.

59 Directorate-General for Research and Innovation, "Options for Strengthening Responsible Research and Innovation," 35 (emphases added).

60 The term "wicked problem" was first defined as a class of planning problems in "open societal systems" that had inadequate theory for forecasting and a "plurality of objectives ... and politics." See Rittel and Webber, "Dilemmas in a General Theory of Planning," 160. Rittel was a founder of the Design Methods Group (DMG) at University of California, Berkeley, during the 1960s, which advocated a rational and scientific approach to urban design. Its emphases were complementary to the work of Richard L. Meier, who joined Berkeley in 1967 and contributed to the DMG *Newsletter*.

Bibliography

ARCHIVAL SOURCES

M. King Hubbert Papers, American Heritage Center (AHC), University of
Wyoming, WY.

Oak Ridge National Laboratory Photo Archives, TN.

Richard Meier Papers, BANC MSS 2008/105, Bancroft Library (BANC),
University of California at Berkeley.

Smithsonian Institution Archives (SIA), Washington, DC.

Technocracy Fonds, University of Alberta Archives (UAA), Edmonton, AB.

Technocracy Fonds, University of British Columbia Archives (UBCA),
Vancouver, BC.

Weinberg Archives, Children's Museum of Oak Ridge (CMOR), TN.

Weinberg Papers, MPA.0332, Modern Political Archives, Howard H. Baker Jr
Center for Public Policy, University of Tennessee, TN.

PUBLISHED SOURCES

Abercrombie, Patrick. "A Civic Society. An Outline of Its Scope, Formation and
Functions." *Town Planning Review* 8, 2 (1920): 79–92.

Abercrombie, Patrick, and Robert H. Matthew. "Clyde Valley Regional Plan 1946."
Edinburgh: His Majesty's Stationery Office, 1949.

Ackerman, Michael. "The Nutritional Enrichment of Flour and Bread: Techno-
logical Fix or Half-Baked Solution." In *The Technological Fix: How People Use
Technology to Create and Solve Problems*, ed. Lisa Rosner, 75–92. New York:
Routledge, 2004.

Adair, David. "The Technocrats 1919–1967: A Case Study of Conflict and Change in a Social Movement." MA thesis, Simon Fraser University, 1970.

Adger, W. Neil. "Social and Ecological Resilience: Are They Related?" *Progress in Human Geography* 24, 3 (2000): 347–64.

Akin, William E. *Technocracy and the American Dream: The Technocracy Movement, 1900–1941*. Oakland: University of California Press, 1977.

Aldiss, Brian W., and David Wingrove. *Trillion Year Spree: The History of Science Fiction*. New York: Scribner, 1973.

Alkon, Paul K. *Science Fiction before 1900: Imagination Discovers Technology*. New York: Routledge, 2002.

American Physical Society Study Group, N. Bloembergen, C.K.N. Patel, P. Avizonis, R.G. Clem, A. Hertzberg, T.H. Johnson, et al. "Report to the American Physical Society of the Study Group on Science and Technology of Directed Energy Weapons." *Reviews of Modern Physics* 59, 3 (1987): S1-S201.

American Red Cross, and International Federation of Red Cross and Red Crescent Societies. "A Vision on the Humanitarian Use of Emerging Technology for Emerging Needs: Strengthening Urban Resilience." New York Red Cross, 2015.

Anderson, Michael. *The Dam Busters*. Associated British Pathe, 1955.

Anderson, Terry H. *The Movement and the Sixties*. Oxford: Oxford University Press, 1996.

Apple Corp. "The Computer for the Rest of Us," television commercial (1930s), 1984.

Armytage, W.H.G. *The Rise of the Technocrats: A Social History*. Milton Keynes: Routledge, 1965.

– *Yesterday's Tomorrows: A Historical Survey of Future Societies*. Toronto: University of Toronto Press, 1968.

Ashley, Mike. *The Gernsback Days: A Study of the Emergence of Modern Science Fiction from 1911 to 1936*. Holicong, PA: Wildside Press, 2004.

Asimov, Isaac. *Foundation*. New York: Gnome Press, 1951.

– *Foundation and Empire*. New York: Gnome Press, 1952.

– *In Memory Yet Green*. New York: Doubleday, 1979.

– *Second Foundation*. New York, 1953.

"Atom Seen Causing 'New Technocracy.'" *New York Times*, 7 January 1946, 17.

Auerbach, Jeffrey. *The Great Exhibition of 1851: A Nation on Display*. New Haven: Yale University Press, 1999.

Augustine, Donna Marie. "Obesity and the Technological Fix: Weight Loss

Surgery in American Women." Master's thesis, Virginia Polytechnic Institute and State University, 2003.

Backus, George, Michael T. Schwein, Scott T. Johnson, and Robert J. Walker. "Comparing Expectations to Actual Events: The Post Mortem of a Y2K Analysis." *System Dynamics Review* 17, 3 (2001): 217–35.

Baker, D. Philip. *The Library Media Program and the School.* Eaglewood, CO: Libraries United Inc., 1984.

Banham, Reyner. *Theory and Design in the First Machine Age.* Cambridge, MA: MIT Press, 1980.

Baran, Paul A. "Review of Meier, Richard L, *Science and Economic Development: New Patterns of Living.*" *American Economic Review* 47, 6 (1956): 1019–21.

Barber, Bernard. "Review of *Reflections on Big Science* by A.M. Weinberg." *American Journal of Sociology* 76, 1 (1970): 169–71.

Barnouw, Erik. *The Golden Web: A History of Broadcasting in the United States, 1933 to 1953.* New York: Oxford University Press, 1968.

Bartlett, Arthur. "Chemical Marvels Take the Bugs Out of Living: From DDT and Other War Developments, Homes Get Promise of Such Miracles as Mosquito Bombs, Wrinkleproof Fabrics and Stickproof Doors." *Popular Science* , May 1945, 150–4.

Bartrip, P.W.J. "History of Asbestos Related Disease." *Postgraduate Medical Journal* 80 (2004): 72–6.

Bashford, Alison , and Philippa Levine, eds. *The Oxford Handbook of the History of Eugenics.* Oxford: Oxford University Press, 2010.

Bauman, Henrikke, and Anne-Marie Tillman. *The Hitchhiker's Guide to LCA: An Orientation on Life Cycle Assessment Methodology and Application.* Lund: Studentlitteratur AB, 2009.

Baxter, Raymond, James Burke, and Michael Latham, eds. *Tomorrow's World.* London: BBC, 1971.

BBC4. *Mad and Bad: 60 Years of Science on TV.* 90 min., 2012.

Beck, Mae. "Steps for Railway and Street Cars." Ed. US Patent Office, 2. USA: Columbia Planograph Co., 1916.

Bélanger, Yves, and Robert Comeau, eds. *Hydro-Québec: Autres Temps, Autres Défis.* Sainte-Foy: Presses de l'Université du Québec, 1995.

Belasco, Warren. "Synthetic Arcadias: Dreams of Meal Pills, Air Food, and Algae Burgers." In *The Technological Fix: How People Use Technology to Create and Solve Problems*, ed. Lisa Rosner, 119–36. New York: Routledge, 2004.

Bell, Michelle L., Devra L. Davis, and Tony Fletcher. "A Retrospective Assessment of Mortality from the London Smog Episode of 1952: The Role of Influenza and Pollution." *Environmental Health Perspectives* 112, 1 (2004): 6–8.

Bell Telephone Inc. *Century 21 Calling*. Seattle World's Fair. https://www.youtube.com/watch?v=liAgTFd9Fo4, 1962.

Bellamy, Edward. *Equality*. New York: D. Appleton & Co., 1897.

– *Looking Backward: 2000–1887*. Boston: Tickner & Co., 1888.

Benintende, Emma Mary. "Who Was the Scientific American? Science, Identity, and Politics through the Lens of a Cold War Periodical." MA thesis, Harvard University, 2011.

Benn, Anthony Wedgewood. "The Government's Policy for Technology." London: Millbank, 1968.

Bentham, Jeremy. *An Introduction to the Principles of Morals and Legislation*. London: T. Payne & Son, 1789.

Berger, Michael *Nano-Society: Pushing the Boundaries of Technology*. RSC Nanoscience and Nanotechnology. Cambridge: Royal Society of Chemistry Publishing, 2009.

Berube, David M. *Nano-Hype: The Truth behind the Nanotechnology Buzz*. New York: Prometheus Books, 2005.

Bethe, Hans. "The Social Responsibilities of Scientists and Engineers." *Cornell Engineer* 29 (1963): 6.

Bijker, Wiebe E. *Of Bicycles, Bakelites, and Bulbs: Toward a Theory of Sociotechnical Change*. Cambridge, MA: MIT Press, 1997.

Bimber, Bruce. "Karl Marx and the Three Faces of Technological Determinism." *Social Studies of Science* 20, 2 (1990): 333–51.

Bitzer, Jürgen, and Phillip J.F. Schröder. "Bug-Fixing and Code-Writing: The Private Provision of Open Source Software." *Information Economics and Policy* 17, 3 (2005): 389–406.

Bompard, E., P. Correia, G. Gross, and M. Amelin. "Congestion-Management Schemes: A Comparative Analysis under a Unified Framework." *IEEE Transactions on Power Systems* 18, 1 (2003): 346–52.

Bookchin, Murray. "Social Ecology versus Deep Ecology: A Challenge for the Ecology Movement." *Socialist Review* 18 (July-Sept. 1988): 9–29.

– "What Is Social Ecology?" In *Environmental Philosophy: From Animal Rights to Radical Ecology*, ed. M.E. Zimmerman, 354–73. Eaglewood Cliffs, NJ: Prentice Hall, 1993.

Bookchin, Murray [under pseudonym Lewis Herber]. *Our Synthetic Environment*. New York: Knopf, 1962.

Boon, Timothy. "'The Televising of Science Is a Process of Television': Establishing *Horizon*, 1962–1967." *British Journal for the History of Science* 48, 1 (2014): 87–121.

Boot, Max. "The New American Way of War." *Foreign Affairs* 82, 4 (2003): 41–58.

Borup, Mads, Nik Brown, Kornelia Konrad, and Harro Van Lente. "The Sociology of Expectations in Science and Technology." *Technology Analysis and Strategic Assessment* 19, 3/4 (2006): 285–98.

Böttcher, Helmuth M. *Wonder Drugs: A History of Antibiotics*. Trans. Einhart Kawerau. Philadelphia: J.B. Lippencott Co., 1964.

Bounds, Leslie. "What's Yours Is Mine." *Technocrat* 14, 3 (1946): 6–9.

Bourassa, Robert. *La Baie James*. Montreal: Éditions du Jour, 1973.

Bowler, Peter. *A History of the Future: Prophets of Progress from H.G. Wells to Isaac Asimov*. Cambridge: Cambridge University Press, 2017.

– *Science for All: The Popularization of Science in Early Twentieth-Century Britain*. Chicago: University of Chicago Press, 2009.

Bradbury, Ray. "Editorial." *Futuria Fantasia* 1, 1 (Summer 1939): 1.

– "Editorial." *Futuria Fantasia* 1, 2 (Fall 1939): 2.

Bradley, Ryan. "The Unbelievable Reality of the Impossible Hyperloop." MIT *Technology Review*, 22 August 2016.

Brady, Joanne. "Electric Car Cultures: An Ethnography of the Everyday Use of Electric Vehicles in the UK." MA thesis, Durham University, 2010.

Brand, Stewart. *The Media Lab: Inventing the Future at MIT*. New York: Viking Penguin, 1987.

– *The Whole Earth Catalog*. Menlo Park, CA: Portola Institute, 1968.

– *Whole Earth Discipline*. London: Atlantic, 2009.

Branscomb, Lewis M. "Harvey Brooks." *Proceedings of the American Philosophical Society* 154, 4 (2004): 461–9.

Brick, Howard. *Transcending Capitalism: Visions of a New Society in Modern American Thought*. Ithaca: Cornell University Press, 2006.

Bridger, Sarah. *Scientists at War: The Ethics of Cold War Weapons Research*. Cambridge, MA: Harvard University Press, 2015.

Bromley, D. Allan. *The President's Scientists: Reminiscences of a White House Science Advisor*. New Haven, CT: Yale University Press, 1994.

Brooks, Fred P. *The Mythical Man-Month*. New York: Addison-Wesley, 1975.

– "No Silver Bullet – Essence and Accidents of Software Engineering." *IEEE Computer* 20, 4 (1987): 10–19.

Brooks, Harvey. "The Evolution of US Science Policy." In *Technology, R&D, and the Economy*, ed. L.R. Bruce and C.E. Barfield, 14–49. Washington, DC: Brookings Institution, 1966.

Brown, Nik, and Mike Michael. "A Sociology of Expectations: Retrospecting Prospects and Prospecting Retrospects." *Technology Analysis and Strategic Management* 15, 1 (2003): 3–18.

Bruce, Robert. "First Planning Report to the Highways and Planning Committee of the Corporation of the City of Glasgow." Glasgow: Corporation of the City of Glasgow, 1945.

Bruening, G., and J.M. Lyons. "The Case of the Flavr Savr Tomato." *California Agriculture* 54, 4 (2000): 6–7.

Buchanan, Nicholas. "The Atomic Meal: The Cold War and Irradiated Foods, 1945–1963." *History and Technology* 21, 2 (2005): 221–49.

Buie, Elizabeth, and Mark Blythe. "Spirituality: There's an App for That! (but Not a Lot of Research)." Paper presented at Computer/Human Interfaces 2013: Changing Perspectives, Paris, France, 2013.

Burchard, John Ely, and James Rhyne Killian. *Q.E.D.: MIT in World War II*. New York: J. Wiley, 1948.

Bureau of Entomology and Plant Hygiene, Agricultural Research Administration. "DDT – for Control of Household Pests." Washington, DC: US Department of Agriculture, US Public Health Service, Federal Security Agency, March 1947.

Burnham, John C. *Accident Prone: A History of Technology, Psychology, and Misfits of the Machine Age* Chicago: University of Chicago Press, 2009.

Burns, Eugene M., and Kenneth E. Studer. "Reflections on Alvin M. Weinberg: A Case Study on the Social Foundations of Science Policy." *Research Policy* 4, 1 (1975): 28–44.

– "Reply to Alvin M. Weinberg." *Research Policy* 5 (1976): 201–2.

Burris, Beverley H. *Technocracy at Work*. Albany, NY: State University of New York Press, 1993.

Bush, Vannevar. *Science: The Endless Frontier*. A Report to the President by Vannevar Bush, Director of the Office of Scientific Research and Development. Washington, DC: United States Printing Office, 1945.

Bylinsky, Gene. "Is 'Big Science' Headed for 'Big Trouble'?" *Huntsville Times*, 15 March 1963.

Byrne, David M., Stephen D. Oliner, and Daniel E. Sichel. "Is the Information Technology Revolution Over?" In *Finance and Economics Discussion Series*. Washington, DC: Divisions of Research and Statistics and Monetary Affairs, Federal Reserve Board, 2013.

Cairncross, Alec K. "Freeing the World from Want: Review of *Science and Economic Development.*" *New Scientist* 25 (9 May 1957): 36–7.

California Air Resources Board. "CARB History." Government of California, 2018. https://ww2.arb.ca.gov/about/history.

Carbon Capture and Storage Association. http://www.ccsassociation.org/.

Carson, Rachel. *The Edge of the Sea.* New York: Houghton Mifflin, 1955.

– *The Sea around Us.* Oxford: Oxford University Press, 1951.

– *Silent Spring.* New York: Houghton Mifflin, 1962.

– *Under the Sea Wind: A Naturalist's Picture of Ocean Life.* New York: Simon and Schuster, 1941.

Casey, Steven M. *Set Phasers on Stun: And Other True Tales of Design, Technology and Human Error.* Santa Barbara, CA: Aegean Publishing Company, 1998.

Cashbaugh, Sean. "A Paradoxical, Discrepant, and Mutant Marxism: Imagining a Radical Science Fiction in the American Popular Front." *Journal for the Study of Radicalism* 10, 1 (2016): 63–106.

Ceruzzi, Paul E. "The 'Problem' of Computer-Computer Communication, 1995–2000: A Technological Fix?" In *The Technological Fix: How People Use Technology to Create and Solve Problems,* ed. Lisa Rosner, 203–18. New York: Routledge, 2004.

Chamon, Marcos, Paolo Mauro, and Yohei Okawa. "Mass Car Ownership in the Emerging Market Giants." *Economic Policy* 23, 54 (2008): 244–96.

Chandler, Arthur. "The First Industrial Exposition." http://www.arthurchandler.com/1798-exposition, reprinted and revised from *World's Fair* 10, 1 (1990).

Chapelle, Francis H. "Bioremediation of Petroleum Hydrocarbon-Contaminated Ground Water: The Perspectives of History and Hydrology." *Groundwater* 37, 1 (2005): 122–32.

Chase, Stuart. *The Tragedy of Waste.* New York: Macmillan, 1925.

CHQ Technocracy Inc. "Symbolization of Technocracy." Report to members, 28 September 1948.

Clarke, Sabine. "Rethinking the Post-War Hegemony of DDT: Insecticides Research and the British Colonial Empire." In *Environment, Health and History,* ed. Virginia Berridge and Martin Gorsky, 133–53. New York: Palgrave Macmillan, 2011.

Club of Rome. *Limits to Growth*. New York: Signet, 1972.

Coates, Peter A. *The Trans-Alaska Pipeline Controversy: Technology, Conservation, and the Frontier*. Cranbury, NJ: Associated University Presses, 1991.

Cochran, Thomas B., Harold A. Feiveson, Walt Patterson, Gennadi Pshakin, M.V. Ramana, Mycle Schneider, Tatsujiro Suzuki, and Frank von Hippel. "Fast Breeder Reactor Programs: History and Status." In *Research Report 8*. International Panel on Fissile Materials, 2010, http://fissilematerials.org/library/rr08.pdf.

Cohen-Cole, Jamie. *The Open Mind: Cold War Politics and the Sciences of Human Nature*. Chicago: University of Chicago Press, 2014.

Cohen, Jerry, and William S. Murphy. *Burn Baby Burn! The Los Angeles Race Riots of August 1965*. London: Avon, 1966.

Coleman, Michel P. "War on Cancer and the Influence of the Medical-Industrial Complex." *Journal of Cancer Policy* 1, 3–4 (2013): e31–e34.

Colignon, Richard A. *Power Plays: Critical Events in the Institutionalism of the Tennessee Valley Authority*. Albany, NY: State University of New York Press, 1997.

Committee on Genetically Engineered Crops. "Genetically Engineered Crops: Experiences and Prospects." Washington, DC: National Academies of Sciences, Engineering, and Medicine (US), 2016.

Conti, Kathleen. "Boston Commute Is as Congested as It Was 10 Years Ago." *Boston Globe*, 2015, https://www.bostonglobe.com/metro/regionals/2015/09/17/zocommute/60AfphVXJRcUJYM4RAFTWK/story.html.

Cottam, Clarence, and Elmer Higgins. "DDT – Its Effect on Fish and Wildlife, Circular-11." In *US GPO*. Washington: US Fish and Wildlife Service, July 1946.

Cotter, Bill. *Seattle's 1962 World's Fair*. Seattle: Arcadia, 2010.

Cottrell, W. F. "Review of Meier, *Science and Economic Development*." *Social Problems* 4, 3 (1957): 261–3.

Cowan, Ruth Schwartz. *A Social History of American Technology*. New York: Oxford University Press, 1997.

Cox, Jim. *American Radio Networks: A History*. New York: McFarland, 2009.

Cross, Nigel, David Elliott, and Robin Roy, eds. *Man-Made Futures: Readings in Society, Technology and Design*. London: Hutchinson, 1974.

Crowther, Geoffrey, William Holford, Oleg Kerensky, Herbert Pollard, T. Dan Smith, and Henry W. Wells. *Traffic in Towns*. London: Her Majesty's Stationery Office, 1963.

Cukier, Kenneth, and Viktor Mayer-Schönberger. "The Dictatorship of Data." *MIT Technology Review* (2013). https://www.technologyreview.com/s/514591/the-dictatorship-of-data/.

Cummiskey, Matthew "There's an App for That: Smartphone Use in Health and Physical Education." *Journal of Physical Education, Recreation and Dance* 82, 8 (2011): 24–30.

"Cure Cancer? Go to Mars? Is Big Science Ruining US Financially?" *Oak Ridger*, 9 May 1961, 1.

Curran, W.J. "The Thalidomide Tragedy in Germany: The End of a Historic Medicolegal Trial." *New England Journal of Medicine* 284 (1971): 481–2.

Cutter, Susan L. "The Landscape of Disaster Resilience Indicators in the USA." *Natural Hazards* 80 (2016): 741–58.

Daunton, Martin, and Bernhard Rieger, eds. *Meanings of Modernity: Britain from the Late-Victorian Era to World War II*. Oxford: Berg, 2001.

Day, Dwayne A., John M. Logsdon, and Brian Latell, eds. *Eye in the Sky: The Story of the Corona Spy Satellites*. Washington, DC: Smithsonian Institution Press, 1998.

De Miranda, Alvaro. "Technological Determinism and Ideology: Questioning the 'Information Society' and the 'Digital Divide.'" In *The Myths of Technology: Innovation and Inequality*, ed. Judith Burnett, Peter Senker, and Kathy Walker, 22–38. New York: Peter Lang, 2009.

De Nevers, Noel, ed. *Technology and Society*. New York: Addison Wesley, 1972.

De Solla Price, Derek. "The Acceleration of Science – Crisis in Our Technological Civilization." *Product Engineering* 32 (1961): 56–9.

– *Little Science, Big Science*. New York: Columbia University Press, 1963.

Del Sesto, Stephen L. "Wasn't the Future of Nuclear Energy Wonderful?" In *Imagining Tomorrow: History, Technology, and the American Future*, ed. by Joseph J. Corn, 58–76. Cambridge MA: MIT Press, 1986.

Dickinson, L.M. *Technocracy Digest*, January 1938, 1.

Diefendorf, Jeffry M., ed. *Rebuilding Europe's Bombed Cities*. New York: Palgrave Macmillan, 1990.

Dikötter, Frank "Race Culture: Recent Perspectives on the History of Eugenics." *American Historical Review* 103, 2 (1998): 467–78.

Directorate-General for Research and Innovation. "Options for Strengthening Responsible Research and Innovation." European Union, 2013.

Dizer, John. *Tom Swift and Company: "Boys' Books" by Stratemeyer and Others*. Jefferson, NC: McFarland, 1982.

Donnelly, Kevin. *Adolphe Quetelet, Social Physics and the Average Men of Science, 1796–1874*. Pittsburgh: University of Pittsburgh Press, 2015.

Dorfman, Joseph. *Thorstein Veblen and His America*. Cambridge, MA: Harvard University Press, 1966.

Drengson, Alan R. *The Practice of Technology: Exploring Technology, Ecophilosophy, and Spiritual Disciplines for Vital Links.* Albany, NY: State University of New York, 1995.

– "The Sacred and the Limits of the Technological Fix." *Zygon* 19, 3 (1984): 259–75.

Drexler, K. Eric. *Engines of Creation: The Coming Era of Nanotechnology.* New York: Doubleday, 1986.

Drucker, Peter. "The Technological Revolution: Notes on the Relationship of Technology, Science, and Culture." *Technology and Culture* 2 (1961): 342–51.

Dubock, Adrian. "The Politics of Golden Rice." GM *Crops and Food* 5, 3 (2014): 210–22.

Dvorsky, George. "Better Living through Transhumanism." *Journal of Evolution and Technology* 19 (2008): 62–6.

Dyer, Davis. "The Limits of Technology Transfer: Civil Systems at TRW, 1965–1975." In *Systems, Experts, and Computers: The Systems Approach in Management and Engineering, World War II and After,* ed. Agatha C. Hughes and Thomas P. Hughes. Cambridge, MA: MIT Press, 2000.

Eden, Amnon H., and James H. Moor. *Singularity Hypotheses: A Scientific and Philosophical Assessment.* Dordrecht: Springer, 2012.

Edgerton, David. *Britain's War Machine: Weapons, Resources and Experts in the Second World War.* London: Allan Lane, 2011.

– *Shock of the Old: Technology and Global History since 1900.* London: Profile Books, 2008.

Egorov, Nikolai N. *The Radiation Legacy of the Soviet Nuclear Complex: An Analytical Overview.* London: Earthscan, 2000.

Eisenhower, Dwight D. "Farewell Address to the Nation." 17 January 1961.

Eisenhower, Dwight D., and Lewis J. Strauss. "A Proposal for Our Time." *Reader's Digest,* June 1968, 75–9.

Ellul, Jacques. *La Technique, Ou L'enjeu Du Siècle.* Paris: Armand Colin, 1954.

– *The Technological Society.* New York: Knopf, 1964.

Elsner Jr, Henry. "Messianic Scientism: Technocracy: 1919–1960." PhD diss., University of Michigan, 1962.

Engels, Friedrich. *Socialism: Utopian and Scientific.* London: Progress Publishers, 1970 [first published as installments in *Revue Socialiste* in 1880].

Etzioni, Amitai, and Richard Remp. *Technological Shortcuts to Social Change.* New York: Russell Sage Foundation, 1972.

European Commission. "A European Strategy for Plastics in a Circular Economy." Brussels, January 2018, http://ec.europa.eu/environment/circular-economy/pdf/plastics-strategy.pdf.

Feld, Bernard T., and Gertrud Weiss Szilard, eds. *The Collected Works of Leo Szilard: Scientific Papers*. Cambridge, MA: MIT Press, 1972.

Finch, Jacqueline. "The Ancient Origins of Prosthetic Medicine." *Lancet* 377, 9765 (2011): 548–9.

Finkbeiner, Ann. *The Jasons: The Secret History of Science's Postwar Elite*. New York: Viking, 2006.

FitzGerald, Frances. *Way out There in the Blue: Reagan and Star Wars and the End of the Cold War*. New York: Simon and Schuster, 2000.

Flader, Susan L. *Thinking Like a Mountain: Aldo Leopold and the Evolution of an Ecological Attitude toward Deer, Wolves, and Forests*. Columbia: University of Missouri Press, 1974.

Fleischer, Richard. *Soylent Green*. Film. 1973.

Foster, Amy. "Futuristic Medievalisms and the US Space Program in Disney's *Man in Space* Trilogy and *Unidentified Flying Oddball*." In *The Disney Middle Ages: A Fairy-Tale and Fantasy Past*, ed. Tison Pugh and Susan Aronstein, 153–67: Palgrave Macmillan, 2012.

Fox, Daniel M. *Engines of Culture: Philanthropy and Art Museums*. New Brunswick: Transaction Publishers, 1963.

Foxon, Timothy J. "Technological Lock-in and the Role of Innovation." In *Handbook of Sustainable Development*, ed. Giles Atkinson, Simon Dietz, and Eric Neumayer, 140–52. Cheltenham: Edward Elgar, 2007.

Franks, Benjamin, Stuart Hanscomb, and Sean F. Johnston. *Environmental Ethics and Behavioural Change*. Abingdon: Routledge Earthscan, 2017.

Frayling, Christopher. *Mad, Bad and Dangerous? The Scientist and the Cinema*. London: Reaktion Books, 2005.

Frenkel, Stephen. "A Hot Idea? Planning a Nuclear Canal in Panama." *Cultural Geographies* 5, 3 (1998): 303–9.

Fry, Shirley A. "Studies of US Radium Dial Workers: An Epidemiological Classic." *Radiation Research* 150, 5 (1998): S21–S29.

Fuller, Buckminster. *Operating Manual for Spaceship Earth*. Carbondale: Southern Illinois University Press, 1968.

Fuller, Buckminster, and Robert W. Marks. *The Dymaxion World of Buckminster Fuller*. New York: Doubleday, 1973.

Fuller, Steve. *Humanity 2.0: What It Means to Be Human Past, Present and Future*. London: Palgrave Macmillan, 2011.

Gabor, Dennis. *Inventing the Future*. London: Secker and Warburg, 1963.

Gabrys, Jennifer. "Plastic and the Work of the Biodegradable." In *Accumulation: The Material Politics of Plastic*, ed. Jennifer Gabrys, Gay Hawkins, and Mike Michael, 208–27. New York: Routledge, 2013.

Galison, Peter, and Barton J. Bernstein. "Physics between War and Peace." In *Science, Technology, and the Military*, ed. E. Mendelsohn, Merritt Roe Smith, and Peter Weingart, 47–86. Dordrecht: Kluwer Academic, 1988.

Galison, Peter, and Bruce William Hevly. *Big Science: The Growth of Large-Scale Research*. Stanford, CA: Stanford University Press, 1992.

Ganz, Cheryl R. *The 1933 Chicago World's Fair: A Century of Progress*. Urbana: University of Illinois Press, 2008.

Geddes, Patrick. "An Analysis of the Principles of Economics." *Proceedings of the Royal Society of Edinburgh* 12 (1884): 943–80.

Gendreau, Paul L., and Frank Vitaro. "The Unbearable Lightness of 'Light' Cigarettes: A Comparison of Smoke Yields in Six Varieties of Canadian 'Light' Cigarettes." *Canadian Journal of Public Health* 96, 3 (2005): 167–72.

Gooday, Graeme. *Domesticating Electricity: Technology, Uncertainty and Gender, 1880–1914*. London: Routledge, 2015.

Gorelik, George. "Bogdanov's Tektology: Its Nature, Development and Influence." *Studies in Soviet Thought* 26 (1983): 39–57.

Greenberg, Daniel S. *Science, Money and Politics: Political Triumph and Ethical Erosion*. Chicago: University of Chicago Press, 2001.

Greenhalgh, Paul. *Ephemeral Vistas: The Expositions Universelles, Great Exhibitions and World's Fairs, 1851–1939*. Manchester: Manchester University Press, 1988.

– *Fair World*. Winterbourne: Papadakis, 2011.

Guffy, Elizabeth E. *Retro: The Culture of Revival*. London: Reaktion Books, 2006.

Haagen-Smit, A.J. "Chemistry and Physiology of Los Angeles Smog." *Industrial and Engineering Chemistry* 44, 6 (1952): 1342–6.

Haber, Heinz. *The Walt Disney Story of Our Friend the Atom*. New York: Simon and Schuster, 1957.

Halley, Richard B., and Harold G. Vatter. "Technology and the Future as History: A Critical Review of Futurism." *Technology and Culture* 19, 1 (1978): 53–82.

Hanna, William, and Joseph Barbera. *The Jetsons*. ABC: Screen Gems, 1962–63.

Hard, Mikael, and Andrew Jamison, eds. *The Intellectual Appropriation of Technology: Discourses on Modernity, 1900–1939*. Cambridge MA: MIT Press, 1998.

Hardin, Garrett. "Lifeboat Ethics: The Case against Helping the Poor." *Psychology Today* 8 (1974): 38–43.

– "The Tragedy of the Commons." *Science* 162, 3859 (1968): 1243–8.

Haring, Kristen. *Ham Radio's Technical Culture*. Cambridge, MA: MIT Press, 2007.

Harrison, Harry. *Make Room! Make Room!* New York: Doubleday, 1966.

Harriss, Joseph. *The Eiffel Tower: Symbol of an Age*. London: Paul Elek, 1975.

Hawisher, Gail E., and Cynthia L. Selfe, eds. *Literacy, Technology, and Society: Confronting the Issues*. Buffalo, NY: Prentice Hall, 1997.

Hawthorn, Tom. "Diehard Few Keep Utopian Dream Alive " *Vancouver Province*, 21 March 1993, A26.

Health and Environment Alliance. "Nanotechnology and Health Risks." HEAL *Fact Sheet* (April 2008).

Heggelund, Gørild *Environment and Resettlement Politics in China: The Three Gorges Project*. London: Routledge, 2004.

Hein, Carola, Jeffry M. Diefendorf, and Yorifusa Ishida, eds. *Rebuilding Urban Japan after 1945*. London: Palgrave Macmillan, 2003.

Herbert, Don. *Mr Wizard's Science Secrets*. USA: Popular Mechanics Co., 1952.

Heumann, Ina. *Gegenstücke: Populäres Wissen Im Transatlantischen Vergleich (1948–1984)*. Wien: Böhlau Verlag, 2014.

Hewlett, Richard G., and Francis Duncan. *Nuclear Navy, 1946–1962*. Chicago: University of Chicago Press, 1974.

Hodges, Andrew. *Alan Turing: The Enigma*. London: Hutchinson, 1983.

Hodnett, Edward. *The Art of Problem Solving; How to Improve Your Methods*. New York: Harper, 1955.

– *The Art of Working with People*. New York: Harper, 1959.

Hopewell, Jefferson, Robert Dvorak, and Edward Kosior. "Plastics Recycling: Challenges and Opportunities." *Philosophical Transactions of the Royal Society B* 364, 1526 (2009): 2115–26.

Horwitch, M. *Clipped Wings: The American SST Conflict*. Cambridge MA: MIT Press, 1982.

Howard, M. Anne, ed. *An Environmental Policy Reader*. Buffalo, NY: Harcourt Brace, 1994.

Høyer, Karl Georg. "The History of Alternative Fuels in Transportation: The Case of Electric and Hybrid Cars." *Utilities Policy* 16, 2 (2007): 63–71.

Hubbert, M. King. "Lesson 22: Industrial Design and Operating Characteristics." In *Technocracy Study Course*, 242–68. New York: Technocracy Inc., 1945.

Hubbert, M. King. *Technocracy Study Course*. New York: Technocracy Inc. 1934.

Hubbert, M. King, and Doel, Ronald. "Oral History Interview M. King Hubbert Session IV, 17 Jan. 1989." American Institute of Physics, https://www.aip.org/history-programs/niels-bohr-library/oral-histories/5031-8.

Hughes, Thomas P. "Afterword." In *The Technological Fix: How People Use Technology to Create and Solve Problems*, ed. Lisa Rosner, 219–40. New York: Routledge, 2004.

– *American Genesis: A Century of Invention and Technological Enthusiasm, 1870–1970*. Chicago: University of Chicago Press, 1989.

Hughes, Thomas Parke. "Technological Momentum." In *Does Technology Drive History? The Dilemma of Technological Determinism*, ed. M.R. Smith and Leo Marx, 101–14. Cambridge, MA: MIT Press, 1998.

Huntington, Ellsworth. *Civilization and Climate*. New Haven, CT: Yale University Press, 1915.

Huntington, Samuel B. *The Soldier and the State*. Cambridge, MA: Belknap Press, 1957.

Huxley, Aldous. *Brave New World*. London: Chatto and Windus, 1932.

– *Letters of Aldous Huxley*. Ed. Grover Smith (New York and Evanston: Harper and Row, 1969).

Huxley, Julian. "Transhumanism." In *New Bottles for New Wines*, ed. Julian Huxley, 13–17. London: Chatto and Windus, 1957.

– *TVA: Adventure in Planning*. London: Architectural Press, 1943.

Ialenti, Vincent. "Adjudicating Deep Time: Revisiting the United States' High-Level Nuclear Waste Repository Project at Yucca Mountain." *Science and Technology Studies* 27, 2 (2014): 27–48.

Inman, Mason. *The Oracle of Oil: A Maverick Geologist's Quest for a Sustainable Future*. New York: W.W. Norton, 2016.

Ivie, Wilton. *America Must Show the Way!* New York: Technocracy Inc., 1938. pamphlet.

Jacobs, Chip, and William J. Kelly. *Smogtown: The Lung-Burning History of Pollution in Los Angeles*. New York: Overlook Press, 2008.

Jacobsen, Annie. *Operation Paperclip: The Secret Intelligence Program That Brought Nazi Scientists to America*. London: Little, Brown and Co., 2014.

Jacobson, Mark Z. *Atmospheric Air Pollution: History, Science and Regulation*. Cambridge: Cambridge University Press, 2002.

Jakle, John A., and Keith A. Sculle. *Lots of Parking: Land Use in a Car Culture*. Charlottesville: University of Virginia Press, 2004.

Jarman, Walter M., and Karlheinz Ballschmiter. "From Coal to DDT: The History of the Development of the Pesticide DDT from Synthetic Dyes Till *Silent Spring*." *Endeavour* 36, 4 (2012): 131–42.

Johnson, Deirdre. *Edward Stratemeyer and the Stratemeyer Syndicate*. New York: Twayne, 1993.

Johnson, Ian. "Technology's Cutting Edge: Futurism and Research in the Red Army, 1917–1937." *Technology and Culture* 59, 3 (2018): 689–718.

Johnson, Martin, Raymond Stokes, and Tobias Arndt. *The Thalidomide Catastrophe: How It Happened, Who Was Responsible and Why the Search for Justice Continues after More Than Six Decades*. Cranbrook: Onwards and Upwards Publishers, 2018.

Johnston, Sean F. "Alvin Weinberg and the Promotion of the Technological Fix." *Technology and Culture* 59, 3 (2018): 620–51.

– *The History of Science: A Beginner's Guide*. Oxford: OneWorld, 2009.

– *Holograms: A Cultural History*. Oxford: Oxford University Press, 2015.

– *Holographic Visions: A History of New Science*. Oxford: Oxford University Press, 2006.

– "Security and the Shaping of Identity for Nuclear Specialists." *History and Technology* 27, 2 (2011): 123–53.

– *The Neutron's Children: Nuclear Engineers and the Shaping of Identity*. Oxford: Oxford University Press, 2012.

– "Technological Fixes as Social Cure-All: Origins and Implications." *IEEE Technology and Society* 37, 1 (2018): 47–54.

– "Technological Parables and Iconic Imagery: American Technocracy and the Rhetoric of the Technological Fix." *History and Technology* 33, 2 (2017): 196–219. "Vaunting the Independent Amateur: *Scientific American* and the Representation of Lay Scientists." *Annals of Science* 75, 2 (2018): 97–119.

Johnston, Sean F., Benjamin Franks, and Sandy Whitelaw. "Crowd-Sourced Science: Societal Engagement, Scientific Authority and Ethical Practice." *Journal of Information Ethics* 26, 1 (2017): 49–65.

Jones, Philip N. "'… A Fairer and Nobler City' – Lutyens and Abercrombie's Plan for the City of Hull 1945." *Planning Perspectives* 13, 3 (1998): 301–16.

Jones, R.V. *Most Secret War: British Scientific Intelligence*. London: Hamish Hamilton, 1978.

Jordan, Andrew, and Timothy O'Riordan. "The Precautionary Principle: A Legal and Policy History." In *The Precautionary Principle: Protecting Public Health, the Environment and the Future of Our Children*, ed. Marco Martuzzi and Joel A. Tickner, 31–48. Copenhagen: World Health Organization, 2004.

Jordan, David R. "Haussmann and Haussmannisation: The Legacy for Paris." *French Historical Studies* 27, 1 (2004): 87–113.

Jordan, John M. *Machine-Age Ideology: Social Engineering and American Liberalism, 1911–1939.* Chapel Hill: University of North Carolina Press, 1994.

Josephson, Paul R. *Would Trotsky Wear a Bluetooth? Technological Utopianism under Socialism, 1917–1989.* Baltimore: Johns Hopkins University Press, 2010.

Jungk, Robert. *Tomorrow Is Already Here: Scenes from a Man-Made World.* London: Rupert Hart-Davis, 1954.

Kabel, Allison, and Catherine Chmidling. "Disaster Prepper: Health, Identity, and American Survivalist Culture." *Human Organization* 73, 3 (2014): 258–66.

Kanigel, Robert. *The One Best Way: Frederick Winslow Taylor and the Enigma of Efficiency.* New York: Viking, 1997.

Kargon, Robert H., Karen Fiss, Morris Low, and Arthur Molella, eds. *World's Fairs on the Eve of War: Science, Technology, and Modernity, 1937–1942.* Pittsburgh: University of Pittsburgh Press, 2015.

Kašpar, Jan, Paulo Fornasiero, and Neal Hickey. "Automotive Catalytic Converters: Current Status and Some Perspectives." *Catalysis Today* 77, 4 (2003): 419–49.

Keating Jr, Joseph C. "Joshua N. Haldeman, Dc: The Canadian Years, 1926–1950." *Journal of the Canadian Chiropractic Association* 39 (1995): 172–86.

Keats, Jonathan. *You Belong to the Universe: Buckminster Fuller and the Future.* New York: Oxford University Press USA, 2016.

Keith, David W. "Geoengineering the Climate: History and Prospect." *Annual Review of Energy and the Environment* 25 (2000): 245–84.

Kelly, Barbara M. *Expanding the American Dream: Building and Rebuilding Levittown.* Albany: State University of New York Press, 1993.

Kelman, Herbert C. "Compliance, Identification, and Internalization: Three Processes of Attitude Change." *Journal of Conflict Resolution* 2, 1 (1958): 51–60.

Kennedy, Paul. *Engineers of Victory: The Problem Solvers Who Turned the Tide in the Second World War.* New York: Random House, 2013.

Kenney, Martin, ed. *Understanding Silicon Valley: The Anatomy of an Entrepreneurial Region.* Stanford, CA: Stanford University Press, 2000.

Kevles, Daniel. *The Physicists: The History of a Scientific Community in Modern America.* New York: Knopf, 1977.

Kimbrell, Ed. "Weinberg-Baker Propose Socio-Techno Institutes." *Knoxville Daily News*, 1967.

King, Benjamin, and Timothy Kutty. *Impact: The History of Germany's V Weapons in World War II.* Cambridge, MA: De Capo, 2009.

Kirby, Maurice W. *Operational Research in War and Peace: The British Experience from the 1930s to 1970.* London: Imperial College Press, 2003.

Kirk, Andrew G. "Appropriating Technology: Alternative Technology, the Whole Earth Catalog and Counterculture Environmental Politics." *Environmental History* 7, 4 (2001): 374–94.

– "Machines of Loving Grace: Appropriate Technology, Environment, and the Counterculture." In *Imagine Nation: The American Counterculture of the 1960s and 1970s*, ed. Michael Doyle and Peter Braunstein, 353–78. New York: Routledge, 2001.

Kirkwood, Adrian, and Linda Price. "Technology-Enhanced Learning and Teaching in Higher Education: What Is 'Enhanced' and How Do We Know? A Critical Literature Review." *Learning, Media and Technology* 39, 1 (2014): 6–36.

Kline, Ronald R., and Trevor Pinch. "Users as Agents of Technological Change: The Social Construction of the Automobile in the Rural United States." *Technology and Culture* 37, 4 (1996): 763–95.

Knight, Damon. *The Futurians.* New York: Gateway, 1977.

Kovarik, W. "Ethyl-Leaded Gasoline: How a Classic Occupational Disease Became an International Public Health Disaster." *International Journal of Occupational Health* 11, 4 (2005): 384–97.

Krige, John. "Atoms for Peace, Scientific Internationalism and Scientific Intelligence." *Osiris* 21 (2006): 161–81.

Lafollette, Marcel C. *Science on American Television: A History.* Chicago: University of Chicago Press, 2013.

Lanouette, William, and with Bela Silard. *Genius in the Shadows: A Biography of Leo Szilard, the Man Behind the Bomb.* New York: Skyhorse, 2013.

Larimer, Robert C. "Treatment of Alcoholism with Antabuse." *Journal of the American Medical Association* 150, 2 (1952): 79–83.

Layton, Edwin. *The Revolt of the Engineers: Social Responsibility and the American Engineering Profession.* Baltimore: Johns Hopkins University Press, 1986.

Leach, Edmund. *A Runaway World?* Tiptree: Anchor Press, 1968.

Lecain, Timothy J. "When Everybody Wins Does the Environment Lose? The Environmental Techno-Fix in Twentieth-Century American Mining." In *The Technological Fix: How People Use Technology to Create and Solve Problems*, ed. Lisa Rosner, 137–54. New York: Routledge, 2004.

Lenk, Hans, ed. *Technokratie Als Ideologie: Sozialphilosophische Beiträge Zu Einem Politischen Dilemma*. Stuttgart: Kohlhammer, 1973.

Lenz, Widukind. "The History of Thalidomide." In *1992 UNITH Conference*: https://www.thalidomide.ca/wp-content/uploads/2017/12/Dr-Lenz-history-of-thalidomide-1992.pdf.

Leopold, Aldo. "The Land Ethic." In *A Sand County Almanac*, 201–26. Oxford: Oxford University Press, 1968.

– *A Sand County Almanac*. Oxford: Oxford University Press, 1968.

Leslie, Stuart W. *The Cold War and American Science: The Military-Industrial-Academic Complex at MIT and Stanford*. New York: Columbia University Press, 1993.

Levitt, I. M. "Review of Meier, *Science and Economic Development*." *Journal of the Franklin Institute* 263, 2 (1957): 170–1.

Lewis, Jonathan E. *Spy Capitalism: Itek and the CIA*. New Haven, CT: Yale University Press, 2002.

Livingston, J.A. "Technocracy Still Lives as Both an Idea and on the Farm." *Philadelphia Evening Bulletin*, 3 August 1967, 16.

L.L.B. "Subsidies and Sabotage." *Northwest Technocrat* 16, 167 (1952): 22.

Lodder, Christina. *Russian Constructivism*. New Haven: Yale University Press, 1983.

London County Council. *The Proud City: A Plan for London*. Film. London: Ministry of Information, 1943.

London Sunday Times, and Elaine Potter. *Suffer the Children: The Story of Thalidomide*. London: Viking Press, 1979.

Loose, Elisabeth. "Maker Culture: Can This Movement Contribute to Environmental Sustainability?" PhD diss., University of Glasgow, forthcoming.

Lownsbrough, John. *The Best Place to Be: Expo 67 and Its Time*. London: Penguin, 2012.

Lucaites, John Louis, and Robert Hariman. "Visual Rhetoric, Photojournalism, and Democratic Public Culture." *Rhetoric Review* 20, 1/2 (2001): 37–42.

Lutyens, Edward, and Patrick Abercrombie. "A Plan for the City and County of Kingston Upon Hull." Hull: A. Brown and Sons, 1945.

Lutz, Catherine, and Anne Lutz Fernandez. *Carjacked: The Culture of the Automobile and Its Effect on Our Lives*. New York: Palgrave Macmillan, 2010.

Lytle, Mark Hamilton. *The Gentle Subversive: Rachel Carson, Silent Spring, and the Rise of the Environmental Movement*. New York: Oxford University Press, 2007.

MacKenzie, Nancy R., ed. *Science and Technology Today*. New York: St Martin's Press, 1995.

Macrae, Stuart. *Winston Churchill's Toyshop*. London: Roundwood Press, 1971.

MacShane, Clay, and Joel A. Tarr. *The Horse in the City: Living Machines in the Nineteenth Century*. Baltimore: Johns Hopkins University Press, 2007.

Maguire, Terry, and David Haslam. *The Obesity Epidemic and Its Management*. London: Pharmaceutical Press, 2009.

Mallet, C.P., ed. *Frozen Food Technology*. London: Blackie, 1993.

Maloney, Mick. "Technocracy Dreams at the Fringe." *Vancouver Courier*, 1 January 1989, 9.

Marchand, Roland. "The Designers Go to the Fair 2: Norman Bel Geddes, the General Motors 'Futurama,' and the Visit to the Factory Transformed." *Design Issues* 8, 2 (1992): 22–40.

Marcuse, Herbert. *One-Dimensional Man: Studies in the Ideology of Advanced Industrial Society*. London: Routledge, 1964.

Marks, Lara V. *Sexual Chemistry: A History of the Contraceptive Pill*. New Haven, CT: Yale University Press, 2010.

Marvin, Carolyn. *When Old Technologies Were New: Thinking about Electric Communication in the Late Nineteenth Century*. Oxford: Oxford University Press, 1988.

Massie, Keith, and Stephen D. Perry. "Hugo Gernsback and Radio Magazines: An Influential Intersection in Broadcast History." *Journal of Radio Studies* 9, 2 (2002): 264–81.

Mattie, Erik. *World's Fairs*. New York: Princeton Architectural Press, 2000.

Mayer, Jorge E. "The Golden Rice Controversy: Useless Science or Unfounded Criticism?" *BioScience* 55, 9 (2005): 726–7.

McConnell, Malcolm. *Challenger: A Major Malfunction*. London: Simon and Schuster, 1987.

McDougall, Walter A. *The Heavens and the Earth: A Political History of the Space Age*. Baltimore: Johns Hopkins University, 1985.

McKellar, Shelley. "Artificial Hearts: A Technological Fix More Monstrous Than Miraculous?" In *The Technological Fix: How People Use Technology to Create and Solve Problems*, ed. Lisa Rosner, 13–30. New York: Routledge, 2004.

McKibben, Bill. *The End of Nature*. London: Viking, 1990.

McNeish, Wallace. "Resisting Colonisation: The Politics of Anti-Roads Protesting." In *Transforming Politics: Power and Resistance*, ed. Paul Bagguley and Jeff Hearn, 67–84. London: Palgrave Macmillan, 1999.

McNichol, Dan. *The Big Dig*. Boston: Silver Lining Books, 2001.

Mead, Margaret. *Blackberry Winter: My Earlier Years*. New York: William Morrow, 1972.

Meier, Richard L. "Automatic and Economic Development." *Bulletin of the Atomic Scientists* 10, 4 (1954): 129–33.

– "A Hopeful Development Path for Africa." *Futures* 28, 4 (1996): 345–58.

– *Modern Science and the Human Fertility Problem*. New York: Wiley, 1959.

– *Planning for an Urban World: The Design of Resource-Conserving Cities*. Cambridge, MA: MIT Press, 1974.

– *Science and Economic Development: New Patterns of Living*. New York: Wiley, 1956.

– "The Social Impact of a Nuplex." *Bulletin of the Atomic Scientists* 25, 3 (1969): 16–21.

– "The World-Wide Prospect." In *Science and Resources: Prospects for Technological Advance*, ed. Henry Jarrett, 139–50. New York: RFF Press, 1959.

Meikle, Jeffrey L. *American Plastic: A Cultural History*. New Brunswick, NJ: Rutgers University Press, 1995.

Menzies, William Cameron. *Things to Come*. Film. London: London Film Co., 1936.

Merton, Robert K. "The Unanticipated Consequences of Purposive Social Action." *American Sociological Review* 1, 6 (1936): 894–904.

Mesthene, Emmanuel G. "On Understanding Change: The Harvard University Program on Technology and Society." *Technology and Culture* 6, 2 (1965): 222–35.

– *Technology and Social Change*. Indianapolis: Bobbs-Merrill Co., 1967.

Michaels, David. "Waiting for the Body Count: Corporate Decision Making and Bladder Cancer in the US Dye Industry." *Medical Anthropology Quarterly* n.s. 2, 3 (1988): 215–32.

Michel, Jean-Baptiste, Yuan Kui Shen, Aviva Presser Aiden, Adrian Veres, Matthew K. Gray, The Google Books Team, Joseph P. Pickett, et al. "Quantitative Analysis of Culture Using Millions of Digitized Books." *Science* 331, 6014 (2011): 176–82.

Michel, John B. "Mutation or Death!" Third Eastern Science Fiction Convention, Philadelphia, 1937.

– "What Is Science Fiction Doing for You?" *Science Fiction Fan* 3, 6 (1939): 7.

Middleton, William D. *The Time of the Trolley*. Milwaukee: Kalmbach Publishing, 1967.

Miles, M. "Disability in an Eastern Religious Context: Historical Perspectives." *Disability and Society* 10, 1 (1995): 49–70.

Mill, John Stuart. *Auguste Comte and Positivism*. USA: Createspace, 2015.

– *Utilitarianism.* London: Parker, Son & Bourn, 1863.

Mirowski, Philip. *More Heat Than Light: Economics as Social Physics, Physics as Nature's Economics.* Cambridge: Cambridge University Press, 1989.

Misa, Thomas J., Philip Brey, and Andrew Feeberg, eds. *Modernity and Technology* Cambridge MA: MIT Press, 2003.

Mohl , Raymond A. "Stop the Road: Freeway Revolts in American Cities." *Journal of Urban History* 30 (2004): 674–706.

Molson, Francis J. "The Boy Inventor in American Series Fiction: 1900–1930." *Journal of Popular Culture* 28, 1 (1994): 31–48.

Mom, Gijs P.A., and David A. Kirsch. "Technologies in Tension: Horses, Electric Trucks, and the Motorization of American Cities, 1900–1925." 42, 3 (2001): 489–518.

Moore, C.J., S.L Moore, M.K. Leecaster, and S.B. Weisberg. "A Comparison of Plastic and Plankton in the North Pacific Central Gyre." *Marine Pollution Bulletin* 42, 12 (2001): 1297–300.

Moore, Gordon E. "Cramming More Components onto Integrated Circuits." *Electronics Magazine*, April 1965, 4.

Moore, Kate. *The Radium Girls, the Dark Story of America's Shining Women.* Naperville, IL: Sourcebooks, 2017.

Morgan, Arthur E. *The Making of the TVA.* Buffalo, NY: Prometheus Books, 1974.

Moriarty, Patrick, and Damon Honnery. "Nuclear Energy: The Ultimate Technological Fix?" In *Rise and Fall of the Carbon Civilisation: Resolving Global Environmental and Resource Problems*, ed. Patrick Moriarty and Damon Honnery, 103–24. London: Springer-Verlag, 2011.

Morozov, Evgeny. *To Save Everything, Click Here: Technology, Solutionism, and the Urge to Fix Problems That Don't Exist.* London: Penguin, 2014.

Morris, William. "Bellamy's Looking Backward." *Commonweal* 5, 180 (22 June 1889).

– *News from Nowhere.* London: Reeves & Turner, 1891.

Morshed, Adnan. "The Aesthetics of Ascension in Norman Bel Geddes's Futurama." *Journal of the Society of Architectural Historians* 63, 1 (2004): 74–97.

Mott, N.F. "The Scientist and Dangerous Thoughts." *Atomic Scientists' News* 2, 7 (1949): 171–2.

Mumford, Lewis. *Technics and Civilization.* San Diego: Harcourt, Brace and World, 1963.

Murphy, Priscilla Coit. *What a Book Can Do: The Publication and Reception of Silent Spring.* Boston: University of Massachusetts Press, 2005.

Nader, Ralph. *Unsafe at Any Speed*. New York: Grossman, 1965.

Naess, Arne. "The Shallow and the Deep, Long-Range Ecology Movement. A Summary." *Inquiry* 16 (1973): 95–100.

Naess, Arne, and Alan Drengson, eds. *The Selected Works of Arne Naess*. Dordrecht: Springer, 2005.

Naess, Arne, and David Rothenberg (trans.). *Ecology, Community, and Lifestyle: Outline of an Ecosophy*. Cambridge: Cambridge University Press, 1993.

"Nanotechnology and the Environment - Hazard Potentials and Risks." *nanowerk*. https://www.nanowerk.com/spotlight/spotid=25937.php.

National Institutes of Health. "Insulin and Human Growth Hormone: Triumphs in Genetic Engineering." *Journal of the American Medical Association* 245, 17 (1981): 1724–5.

Nelkin, Dorothy. *Methadone Maintenance: A Technological Fix*. New York: G. Braziller, 1973.

Neşer, Gökdeniz, Deniz Ünsalan, Nermin Tekoğul, and Frank Stuer-Lauridsen. "The Shipbreaking Industry in Turkey: Environmental, Safety and Health Issues." *Journal of Cleaner Production* 16, 3 (2008): 350–8.

Newman, M.W. "Calls Dread of the Atom Bar to War: Fear Has Kept a Miserable Kind of Peace, Says Scientist." *Chicago Daily News*, 14 June 1960, 18.

Nichols, Joseph L., and George R. Leighton. "Plastics Come of Age." *Harper's Magazine*, June 1942, 300–8.

Nicolia, Alessandro, Alberto Manzo, Fabio Veronesi, and Daniele Rosellini. "An Overview of the Last 10 Years of Genetically Engineered Crop Safety Research." *Critical Reviews in Biotechnology* 34 (2013): 1–12.

Noble, David W. *The Religion of Technology: The Divinity of Man and the Spirit of Invention*. New York: Knopf Doubleday, 1997.

Nolan, William F. *The Ray Bradbury Companion*. Detroit: Gale Group, 1975.

Novak, Matt. "Sunday Funnies Blast Off into the Space Age." In *Smithsonian.com*, 2012. https://www.smithsonianmag.com/history/sunday-funnies-blast-off-into-the-space-age-81559551/.

Novotny, Thomas E., Kristen Lum, Elizabeth Smith, Vivien Wang, and Richard Barnes. "Cigarette Butts and the Case for an Environmental Policy on Hazardous Cigarette Waste." *International Journal of Environmental Research and Public Health* 6 (2009): 1691–705.

Nowlan, Philip Francis. "Armageddon 2419 AD." *Amazing Stories*, August 1928.

Nuclear and Radiation Studies Board. *Lessons Learned from the Fukushima*

Nuclear Accident for Improving Safety and Security of US Nuclear Plants: Phase 2.
National Academies of Sciences, Engineering, and Medicine, Division on Earth
and Life Studies, 2016. https://www.nap.edu/catalog/23488/lessons-learned-
from-the-fukushima-nuclear-accident-two-volume-set.

Nye, David E. *Electrifying America: Social Meanings of a New Technology, 1880–
1940.* Cambridge MA: MIT Press, 1992.

– "The United States and Alternative Energies since 1980: Technological Fix or
Regime Change?" *Theory, Culture and Society* 31, 5 (2014): 103–25.

Oelschlaeger, Max. "The Myth of the Technological Fix." *Southwestern Journal
of Philosophy* 10, 1 (1979): 43–53.

Ojovan, M.I., and W.E. Lee. *An Introduction to Nuclear Waste Immobilisation.*
Amsterdam: Elsevier, 2014.

"Operation Bakersfield: The Impact of Technocracy's Gray Fleet on the Public's
Consciousness Is Mounting Daily. Technocrats to Symbolize at Fair." *Technocrat,*
September 1948, 14–17.

Opler, Ascher. "Fourth-Generation Software." *Datamation* 13 (1967): 22–4.

Oravec, Christine. "Conservationism vs. Preservationism: The 'Public Interest
in the Hetch Hetchy Controversy." *Quarterly Journal of Speech* 70, 4 (1984):
444–58.

Orwell, George. *1984.* London: Secker and Warburg, 1949.

Overington, Michael A. "The Scientific Community as Audience: Toward a
Rhetorical Analysis of Science." *Philosophy and Rhetoric* 10, 3 (1977): 143–64.

Owen, R., P.M. Macnaghten, and J. Stilgoe. "Responsible Research and Innovation: From Science in Society to Science for Society, with Society." *Science and
Public Policy* 39, 6 (2012): 751–76.

Oxo-Biodegradable Plastics Association. "OPA Responds to European Commission: A European Strategy for Plastics in a Circular Economy 16th January 2018
and Report from the Commission to the European Parliament and Its Council
on the Impact of the Use of Oxo-Degradable Plastic, Including Oxo-Degradable
Plastic Carrier Bags, on the Environment 16th January 2018." News release,
January 2018.

Palm, Walter. "Why North America Faces Social Change." *Technocrat* 16, 7 (1948):
7–10.

Parrish, Wayne W. *An Outline of Technocracy.* New York: Farrer and Rinehart, 1933.

Peled, Alon. "Why Did Social Scientists Miss the Bug?" *Computers and Society* 29,
4 (1999): 20–3.

Perkins, John H. "Reshaping Technology in Wartime: The Effect of Military Goals on Entomological Research and Insect-Control Practices." *Technology and Culture* 19 (1978): 169–86.

Perrow, Charles. *Normal Accidents: Living with High Risk Technologies.* Princeton, NJ: Princeton University Press, 1999.

Petkewich, Rachel. "Technology Solutions: Microbes Manufacture Plastic from Food Waste." *Environmental Science and Technology* 37 (2003): 175A–76A.

Phillips, David. "Art for Industry's Sake: Halftone Technology, Mass Photography, and the Social Transformation of American Print Culture 1880–1920." PhD diss., Yale University, 1996.

"Plastics in 1940." *Fortune,* October 1940, 88–95.

Pluvinge, Gonzague. *Expo 58: Between Utopia and Reality.* Tielt: Lannoo, 2008.

Pocock, Rowland Francis. *Nuclear Power: Its Development in the United Kingdom.* London: Unwin Brothers, 1977.

Pogačnik, Anja, and Aleš Črnič. "Ireligion: Religious Elements of the Apple Phenomenon." *Journal of Religion and Popular Culture* 26, 3 (2014): 353–64.

Pohl, Frederik. *The Way the Future Was.* New York: Ballantine, 1978.

Potrykus, Ingo. "Golden Rice and Beyond." *Plant Physiology* 125 (2001): 1157–61.

Powell, David. *Tony Benn: A Political Life.* London: Bloomsbury, 2001.

Preston, Christopher J. *The Synthetic Age: Outdesigning Evolution, Resurrecting Species and Reengineering Our World.* Cambridge, MA: MIT Press, 2018.

Ramo, Simon. *Cure for Chaos: Fresh Solutions to Social Problems through the Systems Approach.* New York: D. Mackay Co., 1969.

Rasmussen, Mikkel Vedby. *The Risk Society at War: Terror, Technology and Strategy in the Twenty-First Century.* Cambridge: Cambridge University Press, 2006.

Raymond, Allen. *What Is Technocracy?* New York: McGraw-Hill, 1933.

Redenbaugh, Keith, Bill Hiatt, Belinda Martineau, Matthew Kramer, Ray Sheehy, Rick Sanders, Cathy Houck, and Don Emlay. *Safety Assessment of Genetically Engineered Fruits and Vegetables: A Case Study of the Flavr Savr Tomato.* New York: CRC Press, 1992.

Reynolds, Ron. "Don't Get Technatal." *Futuria Fantasia* 1, 1 (Summer 1939): 4–5.

Rhees, David J. "A New Voice for Science: Science Service under Edwin E. Slosson, 1921–1929." MA thesis, University of North Carolina at Chapel Hill, 1979.

Rheingold, Howard. *The Virtual Community: Homesteading on the Electronic Frontier.* Cambridge, MA: MIT Press, 2000.

Rhodes, Richard. *The Making of the Atomic Bomb*. London: Simon and Schuster, 1986.

Rieger, Bernhard. "Envisioning the Future: British and German Reactions to the Paris World Fair of 1900." Chap. 7 In *Meanings of Modernity: Britain from the Late-Victorian Era to World War II*, ed. Martin Daunton and Bernhard Rieger, 145–64. Oxford: Berg, 2001.

– *Technology and the Culture of Modernity in Britain and Germany, 1890–1945*. Cambridge: Cambridge University Press, 2005.

Rittel, Horst W.J., and Melvin M. Webber. "Dilemmas in a General Theory of Planning." *Policy Sciences* 4 (1973): 155–9.

Roberts, Alice. *Tamed: Ten Species That Changed Our World*. London: Hutchinson, 2017.

Rodchenko, Alexander. *Experiments for the Future*. New York: Museum of Modern Art, 2005.

Rogers, Heather. *Gone Tomorrow: The Hidden Life of Garbage*. New York: The New Press, 2005.

Rong, Wenge, Zhang Xiong, Dave Cooper, Chao Li, and Hao Sheng. "Smart City Architecture: A Technology Guide for Implementation and Design Challenges." *China Communications* 11, 3 (2014): 56–69.

Rose, Hilary, and Jalna Hanmer. "Women's Liberation, Reproduction and the Technological Fix." In *Sexual Divisions and Society: Process and Change*, ed. Diana Leonard Barker and Sheila Allen, 117–35. London: Tavistock Publications, 1976.

Rosner, Lisa, ed. *The Technological Fix: How People Use Technology to Create and Solve Problems*. New York: Routledge, 2004.

Rostow, Walt Whitman. *The Stages of Economic Growth: A Non-Communist Manifesto*. Cambridge, UK: Cambridge University Press, 1960.

Rowe, Albert P. *One Story of Radar*. Cambridge: Cambridge University Press, 1948.

Rowntree, Harold, and George M. Spencer. "Combined Street-Car Pneumatic Door Device and Brake-Release Mechanism." Ed. US Patent Office, 2. USA: Burdett Rowntree Manufacturing Co, 1910.

Royal Society. "Geoengineering the Climate: Science, Governance and Uncertainty." Royal Society, 2009.

Russell, Dora. *The Religion of the Machine Age*. London: Routledge and Kegan Paul, 1983.

Russo, Enzo, and David Cove. *Genetic Engineering: Dreams and Nightmares.* Oxford: Oxford University Press, 1998.

Ryan, Peter G. "A Brief History of Marine Litter Research." In *Marine Anthropogenic Litter*, ed. Melanie Bergmann, Lars Gutow, and Michael Klages, 1–25. Bern: Springer, 2015.

Rydell, Robert W. *World of Fairs: The Century-of-Progress Expositions.* Chicago: University of Chicago Press, 1993.

Rydell, Robert W., John E. Findling, and Kimberley D. Pelle. *Fair America: World's Fairs in the United States.* Washington, DC: Smithsonian Books.

Salter, Stephen, Graham Sortino, and John Latham. "Sea-Going Hardware for the Cloud Albedo Method of Reversing Global Warming." *Philosophical Transactions of the Royal Society A* 366, 1882 (2008): 3989–4006.

Samuel, Lawrence R. *The End of the Innocence: The 1964–1965 New York World's Fair* Syracuse: Syracuse University Press 2007.

Schiermeier, Quirin. "Iron Seeding Creates Fleeting Carbon Sink in Southern Ocean." *Nature* 428, 6985 (2004): 788.

Schmid, Sonja D. *Producing Power: The Pre-Chernobyl History of the Soviet Nuclear Industry.* Cambridge, MA: MIT Press, 2015.

Schmitt, James Erwin. "From the Frontlines to Silent Spring: DDT and America's War on Insects, 1941–1962." *Concept* 39 (2016): 1–29.

Schrenk, H.H., Harry Heimann, George D. Clayton, W.M. Gafafer, and H. Wexler. "Air Pollution in Donora, Pa. Epidemiology of the Unusual Smog Episode of October 1948. Preliminary Report." In *Public Health Bulletin*, ed. Public Health Service: Federal Security Agency, 1949.

Schrenk, Lisa Diane *Building a Century of Progress: The Architecture of Chicago's 1933–34 World's Fair.* Minneapolis: University of Minnesota Press, 2007.

Schumacher, Ernst F. *Small Is Beautiful: A Study of Economics as If People Mattered.* London: Blond and Briggs, 1973.

Schuman, Leonard M. "Patterns of Smoking Behaviour." In *Research on Smoking Behavior*, ed. Murray E. Jarvik, Joseph W. Cullen, Ellen R. Gritz, Thomas M. Vogt, and Louis Jolyon West, 36–65. National Institute on Drug Abuse Research Monograph, 1977.

Schwartz, Frederic J. *The Werkbund: Design Theory and Mass Culture before the First World War.* New Haven, CT: Yale University Press, 1996.

Sclove, Richard E. "From Alchemy to Atomic War: Frederick Soddy's 'Technology Assessment' of Atomic Energy, 1900–1915." *Science, Technology, and Human Values* 14, 2 (1989): 163–94.

Scott, Howard. "Birthday Talk by Howard Scott." New York, 29 March 1952.

– "'Design, Direction or Disaster': Talk by Howard Scott." 15 March 1958, Detroit.

– "History and Purpose of Technocracy." In *Technocracy – the Design of the North American Technate*. https://archive.org/details/HistoryAndPurposeOfTechnocracy.howardScott Technocracy Inc., 1965.

– "Newspaper Interview." *St Louis Post-Dispatch*, 23 December 1932, 15.

– "Origins of Technical Alliance and Technocracy (Sound Recording, 1963)." https://www.youtube.com/watch?v=r4w4mwCeBWg.

– "Political Schemes in Industry." *One Big Union Monthly*, October 1920, 6–10.

– "Public Lecture by Howard Scott." 1 September 1937, Civic Auditorium, Winnipeg, Manitoba.

– "The Scourge of Politics in a Land of Manna." *One Big Union Monthly*, September 1920, 14–16.

– *The Words and Wisdom of Howard Scott, 1890–1970*. 3 vols. Akron, OH: Section 3, Regional Division 8141, Technocracy Inc., 1989.

Seaborg, Glenn. "Toward a Science of 'Techumology.'" *Air Force and Space Digest* 50, 12 (1967): 82–4.

Seaman, Greg. "Plastics by the Numbers." Eartheasy, https://learn.eartheasy.com/articles/plastics-by-the-numbers/.

Segal, Howard P. *Technological Utopianism in American Culture*. Syracuse: Syracuse University Press, 1985.

Sellers, C. "Discovering Environmental Cancer: Wilhelm Hueper, Post World War II Epidemiology, and the Vanishing Clinician's Eye." *American Journal of Public Health* 87, 11 (1997): 1824–35.

Setlow, Jane K. *Genetic Engineering: Principles and Methods*. New York: Springer, 2007.

Setright, L. J.K. *Drive On! A Social History of the Motor Car*. London: Granta, 2002.

Sheavly, S.B., and K.M. Register. "Marine Debris and Plastics: Environmental Concerns, Sources, Impacts and Solutions." *Journal of Polymers and the Environment* 15, 4 (2007): 301–5.

Sheldrake, John. "Henry Gantt and Humanized Scientific Management." In *Management Theory*, ed. John Sheldrake, 35–43. London: Thompson Learning, 2003.

Showers, Kate B. "Congo River's Grand Inga Hydroelectricity Scheme: Linking Environmental History, Policy and Impact." *Water History* 1, 1 (2009): 31–58.

Silla, Cesare. "Chicago World's Fair of 1893: Marketing the Modern Imaginary of the City and Urban Everyday Life through Representation." *First Monday* 18, 11 (2013), https://firstmonday.org/ojs/index.php/fm/article/view/4955/3787.

Simpson, John W. *Dam! Water, Power, Politics, and Preservation in Hetch Hetchy and Yosemite National Park*. New York: Pantheon, 2005.

Sims, Christo. *Disruptive Fixation: School Reform and the Pitfalls of Techno-Idealism*. Princeton, NJ: Princeton University Press, 2017.

Sitinjak, Efraim, Bevita Meidityawati, Ronny Ichwan, Niken Onggosandojo, and Parinah Aryan. "Enhancing Urban Resilience through Technology and Social Media: Case Study of Urban Jakarta." *Procedia Engineering* 212 (2018): 222–9.

Skinner, H.B.W. "Atomic Energy and the Public Interest." *Discovery* 12, 9 (1951): 269–72.

Slater, Cliff. "General Motors and the Demise of Streetcars." *Transportation Quarterly* 51, 3 (1997): 45–66.

Slovic, Paul, James H. Flynn, and Mark Layman. "Perceived Risk, Trust, and the Politics of Nuclear Waste." *Science* 254, 5038 (1991): 1603–7.

Smalley, Richard E. "Of Chemistry, Love and Nanobots." *Scientific American* 285, 3 (2001): 76–7.

Smith, Alice Kimball. *A Peril and a Hope: The Scientists' Movement in America, 1945–47*. Chicago: University of Chicago Press, 1965.

Smith, Harry. "'Symbolization Drives and Literature Distribution' on West Coast – Operation Golden Gate." *Technocrat*, December 1948, 14–17.

Smith, Jean Edward. *Eisenhower in War and Peace*. New York: Random House, 2012.

Smith, Michael. "Back to the Future: Epcot, Camelot, and the History of Technology." In *New Perspectives on Technology and American Culture*, ed. Bruce Sinclair, 69–81. Philadelphia: American Philosophical Society, 1986.

– *Station X: The Codebreakers of Bletchley Park*. London: Pan, 2000.

Smyth, William M. "Letter to the Editor." *The Living Age*, April 1933, 187.

Soddy, Frederick. *The Role of Money: What It Should Be, Contrasted with What It Has Become*. London: Routledge, 1934.

– *Wealth, Virtual Wealth and Debt*. London: George Allen and Unwin, 1926.

Sorokin, Pitirim. *Contemporary Sociological Theories*. New York: Harper and Brothers, 1928.

Sovacool, Benjamin K., and L.C. Bulan. "Behind an Ambitious Megaproject in Asia: The History and Implications of the Bakun Hydroelectric Dam in Borneo." *Energy Policy* 39, 9 (2011): 4842–9.

Spilhaus, Athelstan. "Oral History Interview by Ronald E. Doel". College Park, MD, USA: Niels Bohr Library and Archives, American Institute of Physics, 1989.

– "Our New Age." Chicago: Hall Syndicate, 26 December 1965.

– "The United States Science Exhibit at the Seattle World's Fair." *Weatherwise* 15, 2 (1962): 47–9.

Stabile, Carol A. *Feminism and the Technological Fix*. Manchester: Manchester University Press, 1994.

Staff Report. "TVA and the Fertilizer Industry." *Agricultural and Food Chemistry* 5, 8 (1957): 570–3.

Staudenmaier, John M. "Recent Trends in the History of Technology." *American Historical Review* 95, 3 (1990): 715–25.

Steffen, Alex. *Worldchanging: A User's Guide for the 21st Century*. New York: Harry N. Abrams, 2008.

Stern, Alexandra Minna. *Eugenic Nation: Faults and Frontiers of Better Breeding in Modern America*. Berkeley: University of California Press, 2005.

Stokes, Elen. "Wanted: Professors of Foresight in Environmental Law!" *Journal of Environmental Law* 31, 1 (2019): 175–86.

Stokes, Raymond G., Roman Köster, and Stephen C. Sambrook. *The Business of Waste: Great Britain and Germany, 1945 to the Present*. Cambridge: Cambridge University Press, 2013.

Storer, Tracy I. "DDT and Wildlife." *Journal of Wildlife Management* 10, 3 (1946): 181–3.

Weinberg, Alvin, and Stephen H. Stow, "An Interview with Alvin Weinberg." Oak Ridge Oral History Project, 2003, http://cdm16107.contentdm.oclc.org/cdm/ref/collection/p15388coll1/id/165.

Strauss, Neil. "Elon Musk: The Architect of Tomorrow." *Rolling Stone*, 15 November 2017.

Strickland, Donald A. *Scientists in Politics: The Atomic Scientists Movement, 1945–46*. Lafayette: Purdue University Studies, 1968.

Sujauddin, Mohammad, Ryu Koide, Takahiro Komatsu, Mohammad Mosharraf Hossain, Chiharu Tokoro, and Shinsuke Murakami. "Characterization of Ship Breaking Industry in Bangladesh." *Journal of Material Cycles and Waste Management* 17, 1 (2015): 72–83.

Taubman, Philip. *Secret Empire: Eisenhower, the CIA and the Hidden Story of America's Space Espionage*. New York: Simon and Schuster, 2003.

Taurog, Norman. *It Happened at the World's Fair*. MGM, 1963.

Taylor, Frederick. *The Berlin Wall: 13 August 1961 – 9 November 1989*. London: Bloomsbury, 2009.

Technical Alliance. *The Technical Alliance: What It Is, and What It Proposes.*
    Pamphlet. New York, 1918.

"Technocracy Demonstration in San Francisco." *Nation* 167, 6 (7 August 1948):
    142–3.

Technocracy Inc. *Operation Columbia.* https://www.youtube.com/watch?v=
    RUregwsi_cw.

– "Operation Ohio Valley." *Technocrat*, December 1950, 23–6.

Teich, Albert H., ed. *Technology and Man's Future.* New York: St Martin's Press,
    1972.

Teles, Steven M. "Kludgeocracy in America." *National Affairs* 36 (2018).
    https://www.nationalaffairs.com/.

Telford, Horace S., and James E. Guthrie. "Transmission of the Toxicity of DDT
    through the Milk of White Rats and Goats." *Science* n.s. 102, 2660 (1945): 647.

Tenner, Edward. *Why Things Bite Back: Technology and the Revenge of Unintended
    Consequences.* New York: Alfred A. Knopf, 1997.

Terzian, Sevan G. "The 1939–1940 New York World's Fair and the Transformation
    of the American Science Extracurriculum." *Science Education* 93, 5 (2009):
    892–914.

– *Science Education and Citizenship: Fairs, Clubs, and Talent Searches for American
    Youth, 1918–1958.* New York: Palgrave Macmillan, 2013.

Thompson, W.B., ed. *Controlling Technology: Contemporary Issues.* Buffalo, NY:
    Prometheus Books, 1991.

Thurston, Alan J. "Paré and Prosthetics: The Early History of Artificial Limbs."
    ANZ Journal of Surgery 77, 12 (2007): 1114–19.

Tilman, Rick. *Thorstein Veblen and His Critics, 1891–1963.* Princeton: Princeton
    University Press, 1992.

Tirosh-Samuelson, Hava. "Technologizing Transcendence: A Critique of Transhu-
    manism." In *Religion and Human Enhancement*, ed. Trothen Tracy J. and Mercer
    Calvin. Cham: Palgrave Macmillan, 2017.

Tobias, Jim. "Technology and Disability." In *The Technological Fix: How People
    Use Technology to Create and Solve Problems*, ed. Lisa Rosner, 61–74. New York:
    Routledge, 2004.

Toffler, Alvin. *Future Shock.* The Bodley Head, 1970.

– *The Third Wave.* New York: Bantam Books, 1980.

Toffler, Alvin, and Heidi Toffler. *Creating a New Civilization: The Politics of the
    Third Wave.* Washington, DC: Progress and Freedom Foundation, 1994.

Tomasson, Richard F. "Review of Meier, *Modern Science and the Human Fertility Problem*." *Social Problems* 8, 3 (1960): 278–9.

Tomiuk, J., K. Wohrmann, and A. Sentker, eds. *Transgenic Organisms: Biological and Social Implications*. Basel: Birkhauser Verlag, 1996.

Trademarkia. "There's an App for That Trademark Information." http://www. trademarkia.com/theres-an-app-for-that-77980556.html.

Trichopoulos, Dimitrios, Anna Kalandidi, Loukas Sparros, and Brian Macmahon. "Lung Cancer and Passive Smoking." *International Journal of Cancer* 27, 1 (1981): 1–4.

Tugwell, Rexford G. "One World – One Wealth." *Ethics: An International Journal of Social, Political and Legal Philosophy* 51, 3 (1951): 173–94.

Turner, Fred. "Buckminster Fuller: A Technocrat for the Counterculture." In *New Views on Buckminster Fuller*, ed. Hsiao-Yun Chu and Roberto G. Trujillo, 146–59. Stanford: Stanford University Press, 2009.

– *From Counterculture to Cyberculture: Stewart Brand, the Whole Earth Network, and the Rise of Digital Utopianism*. Chicago: University of Chicago Press, 2006.

Tutelyan, Victor A., ed. *Genetically Modified Food Sources: Safety Assessment and Control*. Amsterdam: Elsevier, 2013.

Twain, Mark, and Charles Dudley Warner. *The Gilded Age: A Tale of Today*. New York: American Publishing Company, 1873.

Tyldum, Morten. *The Imitation Game*. 114 min. UK: Black Bear Pictures, 2014.

Uekoetter, Frank. "Solving Air Pollution Problems Once and for All: The Potential and the Limits of Technological Fixes." In *The Technological Fix: How People Use Technology to Create and Solve Problems*, ed. Lisa Rosner, 155–74. New York: Routledge, 2004.

United Nations General Assembly. "Treaty on Principles Governing the Activities of States in the Exploration and Use of Outer Space, Including the Moon and Other Celestial Bodies." 1967. http://www.unoosa.org/oosa/en/ourwork/space law/treaties/outerspacetreaty.html.

Urquart, R.N. "Ring out the Old, Ring in the New." *Technocracy Digest* (1945): 23–7.

van Bree, B., G.P.J. Verbong, and G.J. Kramer. "A Multi-Level Perspective on the Introduction of Hydrogen and Battery-Electric Vehicles." *Technological Forecasting and Social Change* 77, 4 (2010): 529–40.

Veblen, Thorstein. *The Engineers and the Price System*. New York: B.W. Huebsch, 1921.

– *The Instinct of Workmanship and the State of the Industrial Arts*. New York: Macmillan, 1914.

– "A Memorandum on a Practical Soviet of Technicians." Chap. 4 In *The Engineers and the Price System*, 86–104. Kitchener: Batoche Books, 2001.

– *Theory of the Leisure Class*. New York: Macmillan, 1899.

Velardo, Joseph T. "Review of *Modern Science and the Human Fertility Problem*." *Journal of Biology and Medicine* 32 (1959): 69–70.

Verbeek, Peter Paul. *Moralizing Technology: Understanding and Designing the Morality of Things*. Chicago: University of Chicago Press, 2011.

Verdoux, Philippe. "Transhumanism, Progress and the Future." *Journal of Evolution and Technology* 20 (2009): 49–69.

Vinge, Vernor. "The Coming Technological Singularity: How to Survive in the Post-Human Era." In *Vision-21: Interdisciplinary Science and Engineering in the Era of Cyberspace*, ed. G.A. Landis, 11–22. Washington, DC: NASA Publication CP-10129, 1993.

"Volger Sees Grim Outlook." *Daily Chronicle* (de Kalb, Illinois), 10 October 1986, 4.

Vollmer, Timothy, and Office for Information Technology. "There's an App for That! Libraries and Mobile Technology: An Introduction to Public Policy Considerations." American Library Association, June 2010.

vom Saal, Frederick S., and Claude Hughes. "An Extensive New Literature Concerning Low-Dose Effects of Bisphenol-a Shows the Need for a New Risk Assessment." *Environmental Health Perspectives* 113, 8 (2005): 926–33.

von Neumann, John. "Can We Survive Technology?" *Fortune* 51 (June 1955): 106.

Vonnegut, Kurt. *Player Piano*. New York: Charles Scribner's Sons, 1952.

Waddington, C.H. *Operational Research in World War II*. London: Harper Collins, 1973.

Wakefield, Ernest H. *History of the Electric Automobile: Battery-Only Cars*. Warrendale, PA: Society of Automotive Engineers, 1993.

Walker, Mick. *Motorcycle: Evolution, Design, Passion*. Baltimore: Johns Hopkins University Press, 2006.

Wang, Jessica. *American Science in an Age of Anxiety: Scientists, Anticommunism, and the Cold War*. Chapel Hill: University of North Carolina Press, 1999.

Wang, Zuoyue. *In Sputnik's Shadow: The President's Science Advisory Committee and Cold War America*. New Brunswick, NJ: Rutgers University Press, 2008.

Warschauer, Mark, and Morgan Ames. "Can One Laptop Per Child Save the World's Poor?" *Journal of International Affairs* 64, 1 (2010): 33–51.

Watt-Smith, Tom. *The Big Life Fix with Simon Reeve*. 60 min: BBC2, 2017–18.

Weart, Spencer R. *The Discovery of Global Warming*. Cambridge, MA: Harvard University Press, 2008.

Weaver, Abi. "Can Technology Strengthen Our Resilience?" International Committee of the Red Cross, 4 March 2015. http://blogs.icrc.org/gphi2/2015/03/04/can-technology-strengthen-our-resilience/.

Weinberg, Alvin M. "Basic Research and National Goals: A Report to the Committee on Science and Astronautics, US House of Representatives, by the National Academy of Sciences." *Minerva* 3, 4 (1965): 499–523.

– "Beyond the Technological Fix." Oak Ridge, TN: Institute for Energy Analysis, Oak Ridge Associated Universities, 1978.

– "'Big Science': A Liability? Excerpts of Talk before the American Rocket Society – ORNL Space-Nuclear Conference." *Christian Science Monitor*, 7 Febrary 1962.

– "Big Science – Marvel or Menace?" *New York Times Magazine*, 23 July 1961, 15, 41, 51.

– "Can Technology Replace Social Engineering?" University of Chicago Alumni Award speech, 11 June 1966.

– "Can Technology Replace Social Engineering?" *Chicago Jewish Forum* 25, 4 (1967): 131–7.

– "Can Technology Replace Social Engineering?" *American Behavioral Scientist* 10, 9 (1967): 710.

– "Can Technology Replace Social Engineering?" *Air Force and Space Digest* 50, 1 (1967): 55–58.

– "Can Technology Replace Social Engineering?" *Bulletin of the Atomic Scientists* 22, 10 (1966): 4–7.

"Can Technology Replace Social Engineering?" *University of Chicago Magazine* 59, 1 (1966): 6–10.

– "Can Technology Replace Social Engineering?" *Scientific Research* 1 (July 1966): 152–6.

– "Can Technology Replace Social Engineering?" *Oak Ridge National Laboratory News*, 17 June 1966.

– "Can Technology Stabilize the World Order?" United Nations Day after-dinner speech, Oak Ridge, TN, 1966.

– "Compassionate Technology: Letter to the Environment Editor." *Saturday Review*, 2 October 1971, 74.

– "A Department of Science, 1985 Edition." *Technology in Society* 8 (1986): 145–7.

– "Edward Shils and the "Governmentalisation" of Science." *Minerva* 34, 1 (1996): 39–43.
– "Effects of Scale on Modern Science and Technology." *Society for the Social Responsibility of Science* 135 (November 1963): 81–6.
– *The First Nuclear Era: The Life and Times of a Technological Fixer*. New York: American Institute of Physics Press, 1994.
– "How Appropriate Is Appropriate Technology?" *Minerva* 18, 3 (1980): 529–31.
– "Impact of Large-Scale Science in the US." *Scientific World* 6, 1 (1962): 8–12.
– "Inherently Safe Technologies (Chemical and Nuclear)." *Institute for Energy Analysis Newsletter* 6, 4 (1984): 3–8.
– "An Interview with Alvin Weinberg." By Stephen H. Stow, Oral History Presentation Program, Oak Ridge, Tennessee, 31 March 2003.
– "Is Technology the Answer to All Our Problems?" *Los Angeles Times*, 9 October 1966, G3.
– "The Limits of Science and Trans-Science." *Interdisciplinary Science Reviews* 2, 4 (1977): 337–42.
– "Nuclear Power and Public Perception." In *Nuclear Reactions: Science and Trans-Science*, 273–89. Washington, DC: American Institute of Physics, 1992.
– *Nuclear Reactions: Science and Trans-Science*. Washington, DC: American Institute of Physics, 1992.
– *Reflections on Big Science*. Boston: MIT Press, 1967.
– "Response to Burns and Studer's 'Reflections on Alvin M. Weinberg.'" *Research Policy* 5 (1976): 197–200.
– "Salvaging the Nuclear Age." *Wilson Quarterly* 3 (Summer 1979): 88–112.
– "The Sanctification of Hiroshima." *Bulletin of the Atomic Scientists* 41, 11 (1985): 34.
– "Social Institutions and Nuclear Energy." *Science* 177 (1972): 27–34.
– "Social Problems and National Socio-Technical Institutes." In *Applied Science and Technological Progress: A Report to the Committee on Science and Astronautics, US House of Representatives, By the National Academy of Sciences*, 415–34, 1967.
– "Technological Fixes and Social Fixes." Commencement address, Stevens Institute of Technology, Hoboken, NJ, 3 June 1973.
– "Technological Fixes, Carbon Dioxide and Water," manuscript, 24 March 1994.
– "Technological Gigantism and the Social Responsibility of the Nuclear Scientist." Presentation before Brookings Institution, Advanced Study Program, 12 February 1964.

– "Technology, Youth and the Environment." Commencement address, Augsburg College, Minneapolis, 31 May 1970.

– "Towards a Compassionate Technology." Commencement address, Worcester Polytechnic Institute, 6 July 1971.

– "A Wake-up Call for Technological Somnambulists: Review of L. Winner, *The Whale and the Reactor*." *Scientist* 1, 4 (1987): 121–2.

– "Will Technology Replace Social Engineering?" Fifteenth Annual Alfred Korzybski Memorial Lecture, 29 April 1966. Harvard Club of New York: Institute of General Semantics, 1966.

"Weinberg Touts 'Technological Fixes' to Stabilize World." *Oak Ridger*, 12 October 1966, 1.

Weinberger, Sharon. *Imaginary Weapons: A Journey through the Pentagon's Scientific Underworld*. New York: Nation Books, 2006.

Weitzel, Elic M., and Brian F. Codding. "Population Growth as a Driver of Initial Domestication in Eastern North America." *Royal Society Open Science* 3, 160319 (2016): http://dx.doi.org/10.1098/rsos.160319.

Welke, Barbara Young. *Recasting American Liberty: Gender, Race, Law, and the Railroad Revolution, 1865–1920*. Cambridge: Cambridge University Press, 2001.

Wellde, Paula T., and Lisa A. Miller. "There's an App for That!" *Journal of Perinatal and Neonatal Nursing* 30, 3 (2016): 198–203.

Weller, Sam. *Listen to the Echoes: The Ray Bradbury Interviews*. New York: Melville House, 2010.

Wells, Christopher W. *Car Country: An Environmental History*. Seattle: University of Washington Press, 2012.

Wells, H.G. *Men Like Gods*. London: Cassell and Co., 1923.

– *A Modern Utopia*. London: Chapman and Hall, 1905.

– *The Outline of History*. London: J.F. Horrabin, 1920.

– *The Shape of Things to Come*. London: Hutchinson, 1933.

– *The Sleeper Wakes*. London 1899 (serial); 1910 (book).

– *The Time Machine*. London: William Heinemann, 1895.

– *The Work, Wealth and Happiness of Mankind*. London: Doubleday, 1932.

– *The World Set Free*. London: Macmillan, 1914.

– "Wanted – Professors of Foresight!" *The Listener* 8 (1932): 729–30.

Westbrok, Robert. "Tribune of the Technostructure: The Popular Economics of Stuart Chase." *American Quarterly* 32 (1980): 387–408.

Westfahl, Gary. *Hugo Gernsback and the Century of Science Fiction*. Jefferson, NC: McFarland and Co., 2007.

"What Is Technocracy?" *Technocrat* 3, 4 (1937): 1.

While, Aidan, Andrew E.G. Jonas, and David Gibbs. "The Environment and the Entrepreneurial City: Searching for the Urban 'Sustainability Fix' in Manchester and Leeds." *International Journal of Urban and Regional Research* 28, 3 (2004): 549–69.

White Jr, Lynn. *Medieval Technology and Social Change*. Oxford: Oxford University Press, 1964.

Whiteley, Nigel. "Toward a Throw-Away Culture: Consumerism, 'Style Obsolescence' and Cultural Theory in the 1950s and 1960s." *Oxford Art Journal* 10, 2 (1987): 3–27.

Whorton, James C. *Before Silent Spring: Pesticides and Public Health in Pre-DDT America*. Princeton, NJ: Princeton University Press, 1979.

Wiener, Lynn Y. "'There's a Great Big Beautiful Tomorrow': Historic Memory and Gender in Walt Disney's Carousel of Progress." *Journal of American Culture* 20, 1 (1997): 111–16.

Wiener, Norbert. *Cybernetics or Control and Communication in the Animal and the Machine*. Cambridge, MA: MIT Press, 1948.

Wigner, Eugene, and Andrew Szanton. *The Recollections of Eugene P. Wigner*. New York: Plenum, 1992.

Wilson, Harold. "Conference address, Labour Party Annual Conference," Scarborough, 1 October 1963.

Wilson, Richard Guy, Diane H. Pilgrim, and Dickran Tashjian. *The Machine Age in America 1918–1941*. New York: Brooklyn Museum/Harry N. Abrams, 1986.

Winner, Langdon. "Do Artifacts Have Politics?" *Daedalus* 109, 1 (1980): 121–36.

Winsor, Dorothy A. *Writing Like an Engineer: A Rhetorical Education*. New York: Routledge, 1996.

Winter, Jerome. "The Michelist Revolution: Technocracy, the Cultural Front, and the Futurian Movement." *Eaton Journal of Archival Research in Science Fiction* 1, 2 (2013): 78–95.

Wirth, Niklaus. "A Brief History of Software Engineering." *IEEE Annals of the History of Computing* 30, 3 (2008): 32–9.

Wise, Herbert. *Breaking the Code*. 75 min: BBC, 1996.

Wolpert, Stuart. "No Quick, Easy Technological Fix for Climate Change." *Hong Kong Engineer* 37, 2 (2009): 14–15.

Wood, Charles H. "The Birth of the Technical Alliance." *New York World*, 20 February 1921.

World Nuclear Association. "Storage and Disposal of Radioactive Waste."
    https://www.world-nuclear.org/information-library/nuclear-fuel-cycle/
    nuclear-waste/storage-and-disposal-of-radioactive-waste.aspx.

Wozniak, Steve, and Gina Smith. *I, Woz: Computer Geek to Cult Icon*. New York:
    W.W. Norton, 2006.

Wunder, Ernst L. "Interrelationship of Smoking to Other Variables and Preventive
    Approaches." In *Research on Smoking Behavior*, ed. Murray E. Jarvik, Joseph
    W. Cullen, Ellen R. Gritz, Thomas M. Vogt, and Louis Jolyon West, 67–95.
    National Institute on Drug Abuse Research Monograph, 1977.

Yerke, Bruce. "The Revolt of the Scientists." *Futuria Fantasia* 1, 1 (Summer 1939):
    2–3.

Yonas, Gerold, and Jill Gibson. *Death Rays and Delusions*. New York: Peter
    Publishing, 2017.

Yousif, Emad, and Raghad Haddad. "Photodegradation and Photostabilization
    of Polymers, Especially Polystyrene: Review." *SpringerPlus* 2 (2013): 398. doi:
    10.1186/2193-1801-2-398.

Index